# Synthese Library

## Studies in Epistemology, Logic, Methodology, and Philosophy of Science

Volume 445

The aim of *Synthese Library* is to provide a forum for the best current work in the methodology and philosophy of science and in epistemology. A wide variety of different approaches have traditionally been represented in the Library, and every effort is made to maintain this variety, not for its own sake, but because we believe that there are many fruitful and illuminating approaches to the philosophy of science and related disciplines.

Special attention is paid to methodological studies which illustrate the interplay of empirical and philosophical viewpoints and to contributions to the formal (logical, set-theoretical, mathematical, information-theoretical, decision-theoretical, etc.) methodology of empirical sciences. Likewise, the applications of logical methods to epistemology as well as philosophically and methodologically relevant studies in logic are strongly encouraged. The emphasis on logic will be tempered by interest in the psychological, historical, and sociological aspects of science.

Besides monographs *Synthese Library* publishes thematically unified anthologies and edited volumes with a well-defined topical focus inside the aim and scope of the book series. The contributions in the volumes are expected to be focused and structurally organized in accordance with the central theme(s), and should be tied together by an extensive editorial introduction or set of introductions if the volume is divided into parts. An extensive bibliography and index are mandatory.

More information about this series at https://link.springer.com/bookseries/6607

Seungbae Park

# Embracing Scientific Realism

Springer

Seungbae Park
School of Liberal Arts
Ulsan National Institute of Science and Technology
Ulsan, Korea (Republic of)

ISSN 0166-6991 ISSN 2542-8292 (electronic)
Synthese Library
ISBN 978-3-030-87812-2 ISBN 978-3-030-87813-9 (eBook)
https://doi.org/10.1007/978-3-030-87813-9



This Springer imprint is published by the registered company Springer Nature Switzerland AG
The registered company address is: Gewerbestrasse 11, 6330 Cham, Switzerland

# Preface

Arthur Fine declared that "the realist programme has degenerated by now to the point where it is quite beyond salvage" (Fine, 1986, p. 149). This declaration was natural in the 1980s. In 1962 and 1980, respectively, Thomas Kuhn and Bas van Fraassen published *The Structure of Scientific Revolutions* and *The Scientific Image* in which they mounted formidable objections to scientific realism. Over the past several decades, however, scientific realists have tackled the forceful objections, criticized scientific antirealism, and advanced new arguments for scientific realism. As a result, scientific realism, in my view, looks better than before. This does not mean, however, that the debate between scientific realists and antirealists has ended. It only means that scientific antirealists have the chance to sharpen their old objections, to develop new objections, to diffuse the criticisms against scientific antirealism, and to refute the new arguments for scientific realism. In my view, the realism debate is as exciting as ever. It is a fruitful research program which constantly provides researchers with enticing new topics.

This book summarizes and develops the ideas that I have already put forward in journals including *Axiomathes*, *Foundations of Science*, *Inquiry*, *Journal of General Philosophy of Science*, *Social Epistemology*, and *Studies in History and Philosophy of Science*. The task of each chapter is as follows. Chapter 1 undermines several standard objections to the no-miracles argument. Chapter 2 criticizes various versions of the pessimistic induction leveled against scientists, realists, and scientific theories. Chapter 3 displays epistemic and pragmatic disadvantages of scientific antirealism. Chapter 4 introduces six new arguments for scientific realism which are distinct from the no-miracles argument. Chapter 5 critically responds to five variants of putative realism. Chapter 6 explores the issue of how to formulate scientific realism and antirealism. Chapter 7 argues that scientific understanding is a boon for scientific realists, but not for scientific antirealists. Chapter 8 develops a new account of scientific progress, and then explores how five rival accounts of scientific progress relate to scientific realism and antirealism. Chapter 9 criticizes mathematical realism and defends mathematical antirealism. Chapter 10 attempts to establish that scientific realism is a better philosophical framework for scientific practice than is scientific antirealism. Chapter 11 presents several new topics for future debates between scientific realists and antirealists.

My deepest intellectual debt goes to my graduate advisor, Richard Healey of the University of Arizona. This book would not exist without the graduate courses that I took from him and the undergraduate courses that I TA-ed for him. The epistemology and philosophy of mind courses that I took from Alvin Goldman also profoundly influenced this book. I also owe an intellectual debt to other great thinkers around the world whose papers and/or books have been inspirational to me. They include Mario Alai, Jamin Asay, Alexander Bird, Benjamin Callard, Finnur Dellsén, Simon Fitzpatrick, Philip Kitcher, Larry Laudan, Peter Lipton, Timothy Lyons, Moti Mizrahi, Alan Musgrave, Thomas Nickles, Hilary Putnam, Stathis Psillos, Samuel Ruhmkorff, Jamie Shaw, Kyle Stanford, Bas van Fraassen, and Brad Wray. My gratitude also goes to anonymous referees of Synthese Library and the journals to which I submitted my papers. Their sharp and insightful comments have been the driving force to improve my thoughts. I am also grateful to my wife (Hyegyoung Gong) and three daughters (Jihye Park, Jeongae Park, and Suzie Park) for the delight that they offered while I was writing this book. I am also thankful to Dave Truncellito and Jill Flohil for helping me on editorial matters. Last but not least, I thank Otávio Bueno, Editor-in-Chief, for inviting me to contribute to Synthese Library.

Ulsan, Korea (Republic of) Seungbae Park

# Contents

**1 In Defense of the No-Miracles Argument** ... 1
1.1 Introduction ... 1
1.2 The NMA ... 2
1.3 Laudan's Objection ... 5
1.4 The Rejection of the Demand ... 7
1.5 IBE ... 8
1.6 Evolutionary Theory ... 11
1.7 The End of History Illusion ... 12
1.8 Ad Hoc ... 13
1.9 Empiricism ... 15
1.10 Surrealism ... 16
1.11 Predictive Similarity ... 19
1.12 Summary ... 20
References ... 20

**2 Critiques of the Pessimistic Induction** ... 23
2.1 Introduction ... 23
2.2 The Pessimistic Induction ... 24
2.3 Problematic Positions ... 26
2.3.1 The Classic PI and IBE ... 26
2.3.2 The Classic PI and Evolutionary Theory ... 27
2.4 The Grand PI ... 28
2.4.1 Four Intrinsic Problems ... 28
2.4.2 The PUA ... 29
2.4.3 The Relative PI ... 32
2.4.4 The PI over Realists ... 33
2.4.5 The Epistemic PI ... 34
2.4.6 The Problem of Misleading Evidence ... 35
2.4.7 Proportional Pessimism ... 40
2.5 Summary ... 41
References ... 42

3 **Critiques of Scientific Antirealism** .......... 45
3.1 Introduction .......... 45
3.2 The PI Over Antirealist Explanations .......... 46
3.3 The PI Over Pessimists .......... 47
3.4 The PI Over Antirealists .......... 48
3.5 Explanation .......... 50
3.6 Prediction .......... 53
3.7 Two Epistemic Disadvantages .......... 54
3.8 Epistemic Reciprocalism .......... 56
3.9 Empiricism and the Problem of Unobserved Anomalies .......... 59
3.10 Constructive Empiricism .......... 60
3.11 Gap Antirealists .......... 62
3.12 Summary .......... 63
References .......... 64

4 **Six New Arguments for Scientific Realism** .......... 67
4.1 Introduction .......... 67
4.2 The Optimistic Induction Over Realists .......... 68
4.3 The Argument from Undiscovered Evidence .......... 69
4.4 The Argument from Neighboring Theories .......... 74
4.5 The Counter-Induction for Realism .......... 76
4.6 The English Notion of Rationality .......... 79
4.7 The Grand Optimistic Induction .......... 83
4.8 Tu Quoque .......... 84
4.9 Summary .......... 85
References .......... 86

5 **Critiques of Five Variants of Putative Realism** .......... 89
5.1 Introduction .......... 89
5.2 Entity Realism .......... 90
5.3 Minimal Realism .......... 93
5.4 Axiological Realism .......... 96
5.5 Van Fraassen's Definition .......... 98
5.6 Selective Realism .......... 101
5.6.1 An Analysis .......... 101
5.6.2 Common Problems .......... 102
5.6.3 Distinctive Problems .......... 105
5.6.4 The PIs against Selective Realism .......... 107
5.6.5 The Six New Arguments .......... 108
5.7 The Argument from a Bad Lot .......... 109
5.8 Summary .......... 111
References .......... 112

6 **Formulating Scientific Realism and Antirealism** .......... 115
6.1 Introduction .......... 115
6.2 Formulational vs. Epistemological .......... 116
6.3 Wide and Narrow Formulations .......... 120

6.4 Epistemic, Attributive, and Conditional Formulations. . . . . . . . . . . 122
6.5 A Pragmatic Advantage of the Epistemic Formulation . . . . . . . . . . 126
6.6 Conceptual and Empirical Objections . . . . . . . . . . . . . . . . . . . . . . . 127
6.7 Doxastic Conditionalism . . . . . . . . . . . . . . . . . . . . . . . . . . . . . . 128
6.8 Acceptance. . . . . . . . . . . . . . . . . . . . . . . . . . . . . . . . . . . . . . 131
6.9 Summary . . . . . . . . . . . . . . . . . . . . . . . . . . . . . . . . . . . . . . 132
References. . . . . . . . . . . . . . . . . . . . . . . . . . . . . . . . . . . . . . . . 133

**7 Scientific Realism and Scientific Understanding** . . . . . . . . . . . . . . . . 135
7.1 Introduction . . . . . . . . . . . . . . . . . . . . . . . . . . . . . . . . . . . . . 135
7.2 No Belief? . . . . . . . . . . . . . . . . . . . . . . . . . . . . . . . . . . . . . . 136
7.2.1 Belief and Understanding . . . . . . . . . . . . . . . . . . . . . . . . 136
7.2.2 Acceptance and Understanding . . . . . . . . . . . . . . . . . . . . . 137
7.2.3 Empiricism and Understanding . . . . . . . . . . . . . . . . . . . . . 140
7.3 No Justification?. . . . . . . . . . . . . . . . . . . . . . . . . . . . . . . . . . . 142
7.3.1 The Episode of Einstein. . . . . . . . . . . . . . . . . . . . . . . . . . 142
7.3.2 Explanation or Prediction . . . . . . . . . . . . . . . . . . . . . . . . 145
7.4 No Truth?. . . . . . . . . . . . . . . . . . . . . . . . . . . . . . . . . . . . . . . 146
7.4.1 Models . . . . . . . . . . . . . . . . . . . . . . . . . . . . . . . . . . . . 146
7.4.2 Past Theories . . . . . . . . . . . . . . . . . . . . . . . . . . . . . . . . 148
7.4.3 Anti-epistemism. . . . . . . . . . . . . . . . . . . . . . . . . . . . . . . 150
7.5 Summary . . . . . . . . . . . . . . . . . . . . . . . . . . . . . . . . . . . . . . 151
References. . . . . . . . . . . . . . . . . . . . . . . . . . . . . . . . . . . . . . . . 152

**8 Scientific Realism and Scientific Progress** . . . . . . . . . . . . . . . . . . . . 155
8.1 Introduction . . . . . . . . . . . . . . . . . . . . . . . . . . . . . . . . . . . . . 155
8.2 The Historical Episode of Semmelweis. . . . . . . . . . . . . . . . . . . . 156
8.3 The Problem-Solving Account. . . . . . . . . . . . . . . . . . . . . . . . . . 158
8.4 The Semantic Account. . . . . . . . . . . . . . . . . . . . . . . . . . . . . . . 159
8.5 The Noetic Account. . . . . . . . . . . . . . . . . . . . . . . . . . . . . . . . 161
8.6 The Epistemic Account . . . . . . . . . . . . . . . . . . . . . . . . . . . . . . 161
8.7 The Evidential Account . . . . . . . . . . . . . . . . . . . . . . . . . . . . . . 163
8.7.1 Content. . . . . . . . . . . . . . . . . . . . . . . . . . . . . . . . . . . . 163
8.7.2 A Comparison with the Problem-Solving Account . . . . . . . 166
8.7.3 A Comparison with the Semantic Account . . . . . . . . . . . . 167
8.7.4 A Comparison with the Noetic Account . . . . . . . . . . . . . . 168
8.7.5 A Comparison with the Epistemic Account. . . . . . . . . . . . 169
8.7.6 Objections and Replies . . . . . . . . . . . . . . . . . . . . . . . . . 170
8.8 Relationships to Realism and Empiricism. . . . . . . . . . . . . . . . . . . 171
8.9 Summary . . . . . . . . . . . . . . . . . . . . . . . . . . . . . . . . . . . . . . 174
References. . . . . . . . . . . . . . . . . . . . . . . . . . . . . . . . . . . . . . . . 175

**9 Scientific Realism and Mathematical Realism**. . . . . . . . . . . . . . . . . . 177
9.1 Introduction . . . . . . . . . . . . . . . . . . . . . . . . . . . . . . . . . . . . . 177
9.2 Mathematical Realism . . . . . . . . . . . . . . . . . . . . . . . . . . . . . . . 178
9.3 The Abstract World . . . . . . . . . . . . . . . . . . . . . . . . . . . . . . . . 179
9.3.1 The Argument from Language. . . . . . . . . . . . . . . . . . . . . . 179

9.3.2 The Queer World ... 180
9.3.3 Tricle ... 181
9.3.4 The Essential Properties of Mathematical Objects ... 182
9.4 The Epistemological Objection ... 183
9.5 Causal Platonism ... 184
9.6 The Indispensability Argument ... 186
9.6.1 Content ... 186
9.6.2 Standard Objections ... 187
9.6.3 The Two Distinctions ... 189
9.6.4 Interactive Theoretical Entities ... 191
9.6.5 Mathematical Convenientism ... 192
9.7 Mathematical Inferentialism ... 193
9.8 Summary ... 197
References ... 198

**10 Scientific Realism and Scientific Practice** ... 201
10.1 Introduction ... 201
10.2 Scientists' Meta-inductions and Historical Inductions ... 202
10.3 The PI and the SI ... 205
10.4 Realism and Scientific Practice ... 208
10.5 Antirealism and Scientific Practice ... 209
10.5.1 Pessimism ... 209
10.5.2 Instrumentalism ... 212
10.5.3 Surrealism ... 214
10.5.4 Noncommittalism ... 216
10.6 Contribution ... 217
10.7 Summary ... 218
References ... 219

**11 New Topics for Future Debates** ... 221
11.1 Introduction ... 221
11.2 Stanford's Two Proposals ... 222
11.2.1 Motivation ... 222
11.2.2 Responses to Stanford's First Proposal ... 223
11.2.3 Responses to Stanford's Second Proposal ... 224
11.3 Future Topics ... 227
11.3.1 Resolving Conflicts ... 227
11.3.2 The Problem of Time Span ... 228
11.3.3 The Problem of Selection ... 230
11.3.4 The Unvirtuous Prediction ... 231
11.3.5 The Problem of Scope ... 232
11.3.6 The Argument from Double Spaces ... 236
11.4 Summary ... 238
References ... 238

**Index** ... 241

# Chapter 1
# In Defense of the No-Miracles Argument

**Abstract** Both semantic realism and epistemic realism inhere in Putnam's no-miracles argument (NMA). Laudan's objection to realism is compatible with both semantic and epistemic realism. To reject the NMA for the reason that the success of science does not cry out for explanation is to violate the basic rule for evaluating an argument. Some antirealists reject inference to the best explanation, but they use it themselves to establish their own philosophical hypotheses. It is incoherent for antirealists to appeal to evolutionary theory and a psychological hypothesis as a way to discredit realism. Realism is not an ad hoc hypothesis because it explains not only the phenomenon that it was designed to explain but also another phenomenon. It is circular to explain success in terms of empirical adequacy. Surrealism inadequately attributes the success of a scientific theory to the behavior of observables. Predictive similarity is an unsatisfactory explanans for explaining success.

**Keywords** Cognitive illusion · Empirical adequacy · Evolutionary theory · Inference to the best explanation · Laudan's objection · No-miracles argument · Predictive similarity · Surrealism

## 1.1 Introduction

There are semantic and epistemic variants of scientific realism in the literature. The semantic version holds that most successful theories are (nearly) true or (approximately) so. The epistemic version holds that successful theories are warranted. Let me call the former "semantic realism" and the latter "epistemic realism" hereafter. The former is so named because it claims that most successful theories have the *semantic* property of being (approximately) true. The latter is so named because it claims that we have sufficient *evidence* for the belief that they are (approximately) true. Throughout this book, "realism" refers to either semantic or epistemic realism, and "antirealism" refers to any position that opposes semantic and epistemic realism.


S. Park, *Embracing Scientific Realism*, Synthese Library 445,
https://doi.org/10.1007/978-3-030-87813-9_1

As Stathis Psillos (1996, p. S306), Howard Sankey (2017, pp. 201–202), and Moti Mizrahi (2020, p. 80) observe, the debate between realists and antirealists has been dominated during the past several decades by two famous arguments: the no-miracles argument (NMA) and the pessimistic induction (PI). The former is viewed as the most forceful argument for realism, and the latter is viewed as the same for antirealism. Although they support their respective opposing positions, they are both intuitively appealing and powerful. The NMA is analyzed and defended in this chapter, and the PI is analyzed and criticized in the next chapter.

This chapter is organized as follows. In Sect. 1.2, I argue that both semantic and epistemic realism inhere in Hilary Putnam's (1975) NMA. In Sect. 1.3, I show, contrary to what many philosophers believe, that Larry Laudan's (1981) objection to realism is compatible with both semantic and epistemic realism. In Sect. 1.4, I claim that to reject the NMA for the reason that the success of science does not cry out for explanation violates the basic rule for evaluating an argument. In Sect. 1.5, I observe that Bas van Fraassen (1989), Laudan (1981), and Arthur Fine (1991) criticize realists for using inference to the best explanation (IBE), but they use it themselves to establish their philosophical hypotheses. In Sect. 1.6, I contend that it is incoherent for van Fraassen (1980) to appeal to Charles Darwin's theory of evolution, a scientific theory, to discredit realism. In Sect. 1.7, I argue that it is self-refuting for Thomas Nickles (2017) to appeal to a psychological hypothesis to discredit realism. In Sect. 1.8, I reply to Greg Frost-Arnold's (2010) objection that realism is an ad hoc hypothesis. In Sect. 1.9, I refute the antirealist proposal that success can be explained by empirical adequacy. In Sect. 1.10, I criticize surrealism, according to which phenomena are as if T, a successful scientific theory, were true. In Sect. 1.11, I expose a few problems with Kyle Stanford's (2000) proposal that the success of T is due to its predictive similarity to its true theory.

## 1.2 The NMA

As noted above, realism asserts that successful theories are (nearly) true or (approximately) so. Two technical terms figure in this formulation of realism. They are "successful" and "approximately true." These two concepts need to be clarified before the NMA is analyzed in this section.

Laudan (1981) has given a clear definition of success. He says that "a theory is successful if it makes substantially correct predictions, if it leads to efficacious interventions in natural order, if it passes a battery of standard tests," i.e., a theory is successful "so long as it has worked well, i.e., so long as it has functioned in a variety of explanatory contexts, has led to confirmed predictions and has been of broad explanatory scope" (Laudan, 1981, p. 23). Stated differently, theories are successful so long as they have high manipulative,[1] explanatory, and predictive powers.

[1] A theory has manipulative power when it helps us intervene in natural processes.

Let me turn to the notion of approximate truth. To say that T is approximately true means that it is false, but it is close to the truth. For example, if it is true that Tom is 190.3 cm, it is approximately true that he is 190 cm. Strictly speaking, it is false that he is 190 cm. However, to say that he is 190 cm is different from saying that he is 90 cm. The former statement is close to the truth, while the latter statement is far from the truth. In other words, the former statement is approximately true, whereas the latter statement is completely false.

It is well known in philosophy of science that it is difficult to precisely define "approximate truth." The term "approximate truth" is vague, like many other predicates in our natural language. We cannot draw a line between approximate truth and complete falsehood any more than we can draw a line between boys and men. It does not follow, however, that vague predicates should not be used. They are usable insofar as they have "clear cases and clear counter-cases" (van Fraassen, 1980, p. 16). In this spirit, I suggest that "approximate truth" is usable insofar as it has "clear cases and counter-cases" (Park, 2014a, p. 272). I have provided such cases above.

The cases also illustrate that an approximately true statement can be more useful than a true statement and that a true statement can be cumbersome. This point applies to science as well. Many scientific theories involve idealized objects, such as frictionless slope. Idealized objects are only similar to their target objects in the world in certain respects. Strictly speaking, the theories invoking idealized objects are false. However, they might be approximately true and more useful than true theories.

The notion of approximate truth is needed not only by realists but also by antirealists. Antirealists believe that observational statements are true. However, some observational statements can only be said to be approximately true (Park, 2014a, p. 272). Moreover, antirealists agree with realists that some scientific theories are successful. Just like approximate truth, however, success comes in degrees, and thus "success" cannot be precisely defined. Therefore, to dismiss "approximate truth" and "success" on the grounds that they are vague is to dismiss the common ground of the debate between realists and antirealists.

Now that we are, to some degree, clear about what realism asserts, we are ready to move on to the issue of what justifies it. Why should we accept realism? Why should we believe that successful theories are (approximately)[2] true? Putnam's answer can be found in the NMA:

> The positive argument for realism is that it is the only philosophy that doesn't make the success of science a miracle. That terms in mature scientific theories typically refer (this formulation is due to Richard Boyd), that the theories accepted in a mature science are typically approximately true, that the same term can refer to the same thing even when it occurs in different theories—these statements are viewed by the scientific realist not as necessary truths but as part of the only scientific explanation of the success of science, and hence as part of any adequate scientific description of science and its relations to its objects. (Putnam, 1975, p. 73)

[2] I drop "approximately" hereafter for the sake of brevity.

In this argument, two hypotheses are considered to explain the success of scientific theories. The first one says that most successful theories are true, which means that no miracle occurs. By contrast, the second one says that they are false, which means that a miracle does occur. Given that a miracle is an unlikely event, the first hypothesis is better than the second one.

I take note of the following three important things about the NMA. First, realism makes the statistical claim that successful theories are *typically* true, i.e., that most successful theories are true. Realism does not make the universal claim that they are true without exception, i.e., that all of them are true (Park, 2018, p. 55). In the next section, readers will see that some antirealists interpret realism as making the universal claim and then provide some obsolete theories that they take to be counterexamples to the universal claim. Such a criticism commits the straw man fallacy.

Second, semantic realism is explicit in the NMA. The NMA claims clearly that most successful theories have the semantic property of being true. Implicit in Putnam's writing quoted above is epistemic realism, according to which successful theories are warranted (Park, 2019a, p. 280). After all, Putnam claims implicitly that realism gives the best explanation for the success of scientific theories, so it is warranted. To say that it is warranted is to say that successful theories are warranted, which is exactly what epistemic realism states. Epistemic realism is similar to Alan Musgrave's definition of realism that "*it is reasonable to believe that* H is true" (Musgrave, 2017, p. 80) and similar to Timothy Lyons's definition of realism that "we can justifiably believe the hypothesis that successful theories are approximately true" (Lyons, 2005, p. 171).

There are several reasons for distinguishing between semantic and epistemic realism (Park, 2019a, pp. 280–283). Let me mention just one of them here. Semantic realism can be true or false, just like a scientific hypothesis, and thus it can be an explanans explaining the success of science. There is no conceptual problem with the explanation that some theories are successful because they are true. By contrast, epistemic realism cannot be true or false, and thus it cannot be an explanans explaining the success of science. It is to put the cart in front of the horse to say that some theories are successful because we can justifiably believe that they are true. We should rather say that we can justifiably believe them because the truth explains their success.

Third, how does the NMA relate to scientists' arguments for their theories? There are two opposing views regarding their relationship that might be called, respectively, "the distinctive view" and "the reductive view." On the distinctive view (Psillos, 2011, p. 188), the NMA is distinct from scientists' arguments. The NMA is a second-order argument, while scientists' arguments are first-order arguments, for successful theories. Scientists' arguments support successful theories, but the NMA additionally supports successful theories. Hence, the NMA has the effect of making successful theories more probable, i.e., were it not for the NMA, successful theories would be less probable.

On the reductive view, by contrast, the NMA is reducible to scientists' arguments for successful theories, i.e., on close examination, "the no-miracles argument is merely a shorthand for all the arguments that scientists provided to support their successful theories" (Park, 2016a, p. 48). The NMA and scientists' arguments may appear to be distinct from each other, but they are numerically identical with each other. The NMA is not a second-order argument, and it does not make successful theories more probable. The support that the NMA provides for successful theories is the same as the support that scientists' arguments provide for successful theories.

Let me motivate this reductive view of the NMA with the use of an analogy. You can give the intensional definition or the extensional definition of "bachelor," i.e., you can say that a bachelor is an unmarried adult male, or you can enumerate the bachelors in the world and then say that they are bachelors. The three terms—"unmarried," "adult," and "male"—are used in the intensional definition, but not in the extensional definition. It is much more convenient to give the intensional definition than to give the extensional definition.

Similarly, you can give the intensional definition or the extensional definition of "realism," i.e., you can say that successful theories are true, or you can enumerate all of the theories in science that have high manipulative, explanatory, and predictive powers, and then say that they are true. The term "successful" is used in the intensional definition, but not in the extensional definition. The enumerated theories, however, commonly have the property of being successful. It is much more convenient to give the intensional definition than to give the extensional definition of "realism" (Park, 2016a, p. 47).

The same holds for the NMA. When requested to justify the truth of the enumerated theories, you can state the NMA, or you can enumerate scientists' arguments for them. When you state the NMA, you are giving the intensional justification for realism. By contrast, when you enumerate scientists' arguments for the enumerated theories, you are giving the extensional justification for realism. Again, it is much more convenient to give the intensional justification than to give the extensional justification (Park, 2016a, p. 49). It is not surprising that Putnam gives the intensional justification instead of the extensional justification for realism.

## 1.3 Laudan's Objection

Laudan (1981) raises an influential objection to realism. The gist of it is that realists' "epistemology is confronted by anomalies which seem beyond its resources to grapple with" (Laudan, 1981, pp. 47–48). By "anomalies," he means false theories that were successful in the history of science. He specifies them in his infamous list of obsolete theories:

**Laudan's List**

- the crystalline spheres of ancient and medieval astronomy;
- the humoral theory of medicine;
- the effluvial theory of static electricity;
- "catastrophist" geology, with its commitment to a universal (Noachian) deluge;
- the phlogiston theory of chemistry;
- the caloric theory of heat;

- the vibratory theory of heat;
- the vital force theories of physiology;
- the electromagnetic aether;
- the optical aether;
- the theory of circular inertia;
- theories of spontaneous generation (Laudan, 1981, p. 33).

Laudan takes these theories to be anomalies to realism, which indicates that he takes realism as making the universal claim that *all* successful theories are true, but not the statistical claim that *most* successful theories are true (Park, 2011, pp. 22–23).

Lyons (2016, 2017, 2018) advances the same objection to realism. However, he formulates it in terms of modus tollens:

1. If (a) that realist meta-hypothesis were true, then (b) we would have no successful theories that cannot be approximately true. (If we did, each would be a "miracle," which no one of us accepts.)
2. However, (not-b) we do have successful theories that cannot be approximately true: the list (of "miracles").
3. Therefore, (not-a) the realist meta-hypothesis is false. (And the no-miracles argument put forward to justify that meta-hypothesis is unacceptable.) (Lyons, 2016, p. 566).

Lyons' objection to realism looks different from Laudan's, but it is, in fact, the same. Both are making the same point that past theories make it false that all successful theories are true (Park, 2019b, p. 608). Consequently, I call the two objections "Laudan's objection" from now on.

On close examination, Laudan's objection refutes neither semantic realism nor epistemic realism. Semantic realism claims that most, not all, successful theories are true, as noted in Sect. 1.2. Consequently, it is compatible with the existence of the obsolete theories on Laudan's list (Mizrahi, 2013, p. 3224; Park, 2019b, p. 607). Epistemic realism, which asserts that successful theories are warranted, is also compatible with Laudan's objection. According to epistemologists, a justified belief is not guaranteed to be true but rather is likely to be true. Alvin Goldman and Bob Beddor say, for instance, "The truth-ratio need not be 1.0, but the threshold must surely be greater (presumably quite a bit greater) than .50" (Goldman & Beddor, 2016). Thus, epistemic realism is compatible with Laudan's list.

Lyons anticipates this defense of realism and objects that realism concedes "to miracles" (Lyons, 2016, p. 572). His point is that if realism makes the statistical claim, it allows for some false theories to be successful, which means that realism allows for some miracles to occur. Given that miracles are unlikely events, it is unreasonable to invoke them to explain the success of false theories.

This preemptive objection is built upon the interpretation of the NMA that the number of miracles coincides with the number of successful but false theories. This interpretation is disputable. Putnam states that realism "is the only philosophy that doesn't make the success of science a miracle" (Putnam, 1975, p. 73). His statement indicates that he regards the success of science, not the success of a single theory, as a miracle. It is my interpretation that according to the NMA, a miracle happens not when a single false theory is successful but rather when realism is false, i.e., when most successful theories are false (Park, 2019b, p. 610).

## 1.4 The Rejection of the Demand

Van Fraassen contends that observable regularities do not need to be explained, saying "that the observable phenomena exhibit these regularities, because of which they fit the theory, is merely a brute fact, and may or may not have an explanation in terms of unobservable facts 'behind the phenomena'—it really does not matter to the goodness of the theory, nor to our understanding of the world" (van Fraassen, 1980, p. 24). Robin Hendry also suggests that instrumentalists "typically reject the call for explanation" (Hendry, 2001, p. S25). In the vein of these philosophers, some antirealists might reject the NMA, claiming that the success of science does not cry out for explanation.

How can realists respond to this criticism of the NMA? Suppose that antirealists explain something, whatever it is. For example, they explain the existence of the fossils of sea life on Mt. Everest in terms of plate tectonics. Realists can challenge antirealists to explain why it is that the fossils call for explanation whereas the success of science does not (Park, 2011, p. 26). It is worth noting that while van Fraassen claims that observable regularities do not need to be explained, he explains other phenomena. For example, he explains why science is successful (van Fraassen, 1980, pp. 39–40) and why the explanatory phenomena called "rejections" and "asymmetries" occur in science (van Fraassen, 1980, p. 112). Therefore, realists can challenge him to explain why it is that the success of science and the explanatory phenomena long for explanations, whereas observable regularities do not.

What if antirealists say that nothing pines for explanation and they do not explain anything? In response, realists can say that antirealists and realists merely have different tastes, i.e., that realists are interested, whereas antirealists are not, in explaining the success of science. Moreover, antirealists' lack of motivation to explain the success of science does not negatively affect the adequacy of realists' explanation of it (Park, 2011, p. 26). In other words, realists' explanation of it is adequate or inadequate independently of whether antirealists are motivated to explain it. Therefore, realists do not need to motivate antirealists to explain it.

In addition, imagine that you construct an argument, whatever it is, and that I dismiss it on the grounds that I am not motivated to construct it. You would point out that my rejection of your argument violates the basic rule of how to attack an argument. The basic rule is that we should attack either a premise or the inference from the premises to the conclusion. Consequently, I should attack either a premise or the form of your argument, instead of dismissing your argument on the grounds that I am not motivated to construct it.

Realists can say the same thing to antirealists. Antirealists reject the NMA on the grounds that they are not motivated to explain the success of theories. Realists can point out that antirealists violate the basic rule of how to attack an argument and that antirealists should attack either the explanandum or the inference from the explanandum to the explanans of the NMA, instead of saying that they are not motivated to give the realist explanation, i.e., to construct the NMA.

The foregoing criticisms apply no less to antirealists who claim that observable regularities do not need to be explained. Scientists explain observable regularities with the use of scientific theories. Antirealists say that they are not motivated to explain observable regularities. Under these circumstances, scientists can say to antirealists that they merely have different tastes. Antirealists' motivational inertness to explain observable regularities does not negatively affect the adequacy of their explanations of observable regularities. To reject their explanations on the grounds that observable regularities do not long for explanation is to violate the basic rule of how to attack an argument. Consequently, they are not motivated to motivate antirealists to explain observable regularities.

## 1.5 IBE

We use IBE when we believe that a hypothesis that best explains phenomena is true. Philosophers and scientists rely on IBE to establish their hypotheses. For example, Darwin (1859/1993, *passim*) uses IBE to argue for evolutionary theory. He says, "These are strange relations on the view of each species having been independently created, but are intelligible if all species first existed as varieties" (Darwin, 1859/1993, p. 202). Similarly, Putnam (1975) states that the success of science is intelligible on the view that successful theories are true, but unintelligible on the view that a miracle occurs.

Van Fraassen (1989) constructs an influential argument called "the argument from a bad lot" to attack IBE. It is vigorously defended by James Ladyman et al. (1997), Brad Wray (2008, p. 322, 2012, p. 377), and Kareem Khalifa (2010, pp. 95–97). Van Fraassen puts it as follows:

> So our selection may well be the best of a bad lot. … For me to take it that the best of set X will be more likely to be true than not, requires a prior belief that the truth is already more likely to be found in X, than not (van Fraassen, 1989, p. 143).

Before believing the best of conceived hypotheses, realists need to establish that a true hypothesis is more likely to fall in the collection of considered hypotheses than in the collection of unconsidered hypotheses. To put it differently, they need to show that scientists "are by nature predisposed to hit on the right range of hypotheses" (van Fraassen, 1989, p. 143) or that "none of the possible explanations we have failed to come up with is as good as the best of the ones we have" (Ladyman et al., 1997, p. 306). Consequently, realists would have to compare the best of conceived explanations with unconceived ones.

There is an intrinsic flaw with the argument from a bad lot. The premise is a possibility statement that there *might* be unconceived alternatives, whereas the conclusion is a *normative* statement that realists need to establish that a true hypothesis is more likely to fall in the collection of imagined hypotheses than in the collection of unimagined hypotheses. An inference from a descriptive statement to a normative statement is fallacious (Hume, 1888/1978, p. 469). Therefore, an inference from a possibility statement to a normative statement is a fortiori fallacious (Park, 2017a, pp. 30–31).

Van Fraassen might develop a theory of warrant to show that the best of the conceived hypotheses is unwarranted. If he did, however, his critics would apply the argument from a bad lot to his theory of warrant, saying that there are two groups of theories of warrant. One is a group of conceived ones, and the other is a group of unconceived ones. Before believing the best of the first group, van Fraassen needs to establish that the correct theory is more likely to fall in the first group than in the second group. Thus, he would have to compare the best of the conceived theories with the unconceived ones (Park, 2017a, pp. 30–31).

I grant, however, that van Fraassen's critiques of IBE, including the argument from a bad lot, are all correct. In this book, I am more interested in elucidating the disastrous implications of rejecting IBE for those antirealists who reject IBE but use IBE themselves to establish their own philosophical hypotheses. Those antirealists include van Fraassen, Laudan, and Fine. Let me discuss these philosophers' appeals to IBE one by one.

Van Fraassen (1980) uses IBE to justify his theory of explanation called "the contextual theory" (Park, 2017b, pp. 61–62). It holds roughly that "An explanation is an answer to a why-question" (van Fraassen, 1980, p. 134), and "among the scientifically relevant factors, context determines explanatorily relevant ones" (van Fraassen, 1980, p. 126). How does he argue for his theory? He asserts that the correct theory of scientific explanation should meet the following condition:

> To be successful, a theory of explanation must accommodate, and account for, both rejections and asymmetries. I shall now examine some attempts to come to terms with these, and gather from them the clues to the correct account (van Fraassen, 1980, p. 112).

It does not matter for the purpose of this book what rejections and asymmetries are. What matters is that van Fraassen (1980, p. 111, pp. 132–134) claims that the contextual theory can explain rejections and asymmetries, while its competitors cannot, so the former is correct and the latter are incorrect. Note that he uses IBE to establish his philosophical theory, just as scientists do to establish their scientific theories. Therefore, van Fraassen has to reject either the contextual theory or his critiques of IBE, including the argument from a bad lot.

The second antirealist whom I discuss is Laudan. In the context of criticizing the NMA, he says that the NMA is "a monumental case of begging the question" (Laudan, 1981, p. 45). His point is that some antirealists reject IBE, but the NMA exemplifies IBE, so the NMA would not be persuasive to them. My response to Laudan is to point out that Laudan (1984) uses IBE himself. He says, "Science is successful, to the extent it is successful, because scientific theories result from a winnowing process which is arguably more robust and more discriminating than other techniques we have found for checking our empirical conjectures about the physical world" (Laudan, 1984, p. 101). Note that he explains the success of science in terms of the winnowing process, which implies that he uses IBE to explain the success of science. It follows that his objection to the NMA applies no less to his own proposal.

The third antirealist whom I discuss is Fine. In the context of criticizing the NMA, Fine says, like Laudan, that the NMA is "a paradigm case of begging the question" (Fine, 1991, p. 82). My reply to this objection is that Fine (1991) uses IBE himself to explain the success of theories. He says that to "explain instrumental success we need only suppose that our hypotheses and theories are instrumentally reliable" (Fine, 1991, p. 83). To say that theories are instrumentally reliable means that they have the disposition to be successful:

> This is an explanation of outcomes by reference to inputs that have the capacity (or "power") to produce such outcomes. It is a typical dispositional explanation, just like Newton's account of the phenomena of gravitation (Fine, 1991, p. 83).

Thus, Fine's explanation is that T is successful because it was disposed to be successful. A problem with this explanation is that it has the form that something does X because it is disposed to do X and that the explanatory schema can be applied to anything, casting no light on why X occurs (Park, 2015a, p. 92). However, I set this intrinsic problem aside here. Instead, I point out that Fine uses IBE to give his own explanation of the success of T, so his objection to the NMA is an example of being hoisted on one's own petard.

Fine might reply that IBE is reliable when it concerns only observables, although it is not reliable when it concerns unobservables. However, this reply faces the following counters. First, it is not clear whether his explanatory property, disposition, is observational or theoretical. Second, it is not clear why IBE is reliable when it concerns only observables, but not when it concerns unobservables; there should be a justification for the different attitudes toward the different uses of IBE. It is useful to note here that Ladyman et al. reject the different attitudes, saying, "the rule of IBE is unacceptable in general" (Ladyman et al., 1997, p. 312). On Ladyman et al.'s account, IBE is unacceptable whether inference is made to observables or unobservables.

Fine and Laudan might retort that they do not use IBE when they account for the success of science, given that they do not contend that their accounts are better than the realist account. They are only claiming that there are antirealist alternatives to the realist account and that realists have the burden of proving that their account is better than the antirealist alternatives. Consequently, it is wrong for me to accuse them of holding a double standard regarding their explanations and the realist explanation.

This retort motivates me to distinguish between IBE and inference to explanation. We use inference to explanation when we believe T on the grounds that it explains phenomena. We neither claim nor make an effort to show that it is better than its competitors. My objection to inference to explanation is that if IBE is dubious, inference to explanation is a fortiori dubious, and thus Laudan and Fine cannot escape from my objection by saying that they do not claim that their explanations are superior to the realist explanation (Park, 2015b, p. 27).

Van Fraassen, Laudan, and Fine are in a dilemma. They can either accept or reject IBE. On the one hand, if they accept IBE, they can accept their philosophical hypotheses, but they would also have to accept scientific hypotheses, which amounts to giving up antirealism. On the other hand, if they reject IBE, they can stick to antirealism, but they would have to reject their philosophical hypotheses. The bottom line is that they cannot have it both ways, viz., rejecting IBE and accepting their philosophical hypotheses.

## 1.6 Evolutionary Theory

Van Fraassen (1980, pp. 39–40) advances an antirealist alternative to the realist explanation of the success of science. Surprisingly, he does not utilize the concept that he developed, viz., empirical adequacy. He instead appeals to evolutionary theory:

> For any scientific theory is born into a life of fierce competition, a jungle red in tooth and claw. Only the successful theories survive—the ones which *in fact* latched on to actual regularities in nature (van Fraassen, 1980, p. 40).

This explanation is allegedly an antirealist competitor to the realist explanation because it invokes not the truth but rather the selection of successful theories. With such an explanation, antirealists can allegedly meet the demand to explain the success of science, while remaining uncommitted to the truth of successful theories.

How do I respond to the evolutionary explanation? As we noted in Sect. 1.5, Darwin (1859/1993, *passim*) relies on IBE to support evolutionary theory. Van Fraassen claims, however, that IBE "is unacceptable in general" (Ladyman et al., 1997: 312). In other words, IBE is unacceptable whether the best explanation invokes observables or unobservables. This claim about IBE implies that van Fraassen does not even believe that evolutionary theory is empirically adequate (EA). Therefore, he cannot say, "Only the successful theories survive—the ones which *in fact* latched on to actual regularities in nature" (van Fraassen, 1980, p. 40). To say so and to reject IBE in toto is to try to have it both ways.

There is another problem with van Fraassen's application of evolutionary theory to science. Constructive empiricism claims that science "aims to give us theories which are empirically adequate" (van Fraassen, 1980, p. 12). Note that according to constructive empiricism, science has an aim. What is wrong with such a view of science? Natural selection is a process that is free of a goal, and species do not evolve toward a goal. Therefore, philosophers who apply evolutionary theory to science have to reject the view that science has an aim (Park, 2017c, 328).

Like van Fraassen, Wray (2007, 2010) defends the evolutionary explanation. He says that "selection mechanism operative in science, like natural selection in the biological world, is essentially an eliminative process" (Wray, 2010, p. 376). Unlike van Fraassen, however, Wray (2013, 2015) defends the PI. In my view, it is incoherent to defend both the evolutionary explanation *and* the PI. The PI implies that evolutionary theory will run into anomalies, just as its predecessors ran into anomalies, and hence that evolutionary theory is not EA, just as its forerunners were not EA (Park, 2001, p. 78; Lyons, 2003, p. 898). Consequently, the evolutionary explanation is not EA.

Van Fraassen and Wray have made a mistake similar to the one made by Kathleen Akins (1996). She appeals to evolutionary theory to justify her view that our perceptual system produces not veridical information but rather useful information, the information of which our motor system can make use. She says, "For every evolved system, there will be a symbiotic relationship between the information gathering of the sensory system and the informational needs of the motor system—and the elegant solutions that evolution eventually selects need not involve any straightforward (to our eyes) 'veridical' encoding of sensory information" (Akins, 1996, p. 353). This argument is self-undermining (Park, 2017c, p. 330). After all, Darwin perceived the world to gather evidence for evolutionary theory. If Akins were right, however, the evidence that Darwin gathered would not be veridical but rather merely useful. It is arbitrary to suggest that Darwin's perceptual system was special, so the information that he gathered was not merely useful but rather veridical.

## 1.7 The End of History Illusion

A toddler's life is radically different from a 10-year-old child's life, which is, in turn, radically different from a 20-year-old college student's life, which is, in turn, radically different from a 30-year-old office worker's life, and so on. As we grow older, our preferences and interests go through radical transformations. However, we tend to believe that we will not go through such transformations in the future, although we believe that we have gone through such transformations in the past. To state it differently, we tend to believe that there will be no dramatic changes in our future, although we believe that there were many such changes in our past. This cognitive illusion is called "the end of history illusion" (Quoidbach et al., 2013, p. 98). We are under the illusion that the present is the moment when eventful lives end and uneventful lives begin.

Psychologists postulate the existence of this cognitive illusion on the basis of a psychological experiment. In the experiment, a group of subjects are asked to assess their lives over the past 10 years. They state that their lives have been dynamic. Another group of subjects are asked to assess their lives over the next 10 years. They state that their lives will not be dynamic. Their different estimations indicate that "people underestimate the extent to which their personalities, values, and preferences will change in the future" (Quoidbach et al., 2013, p. 98). Note that psychologists postulate the existence of the cognitive illusion to explain the conflicting reports of the two groups of subjects. Their psychological hypothesis says that people are under the end of history illusion, which explains why the two groups of subjects give the radically different estimations.

Nickles (2017) invokes this psychological hypothesis to discredit realism. Realism asserts that there will be few scientific revolutions, although there were many scientific revolutions in the past. Realists "see the future of a currently successful field as similarly flat—as if the dynamical, innovative history of the field is now ended" (Nickles, 2017, p. 152). They "cannot fully resist the temptations of the cognitive-historical illusions" (Nickles, 2017, p. 162). Stated differently, realism originates from the end of history illusion. Therefore, it is not worthy of our beliefs.

What are we to make of Nickles's case against realism? It is not clear whether realism was caused by the end of history illusion, although there is a similarity between them. As far as I know, no realist appeals to the illusion to justify realism in the literature. Moreover, realists can retort that even if realism is caused by the cognitive illusion, it is entirely a separate issue whether realism is tenable. We should accept or reject realism not on the basis of whether it is caused by a cognitive illusion but rather on the basis of whether the argument for it is strong or weak. To reject realism on the grounds that it is caused by a cognitive illusion commits a genetic fallacy (Park, 2019c, p. 63). A genetic fallacy arises when we take the cause of a belief as the evidence for or against the belief.

In addition, it is self-refuting to appeal to a psychological hypothesis to discredit realism. The psychological hypothesis is a scientific hypothesis. It follows that appealing to it to discredit realism is just as incoherent as appealing to evolutionary theory to discredit realism. Consider the following dilemma against Nickles. The psychological hypothesis is either true or false. On the one hand, if Nickles believes that it is true, he should also believe that other scientific hypotheses are true, given that they enjoy more or less the same degree of success. If he believes scientific hypotheses, however, he would be a realist. On the other hand, if he believes that it is false, he should also believe that his appeal to it is epistemically impotent, given that an appeal to a false hypothesis does not prove anything. Therefore, he should either be a realist or believe that he does not have a case against realism.

In addition, Nickles (2017, p. 159) rejects the NMA on the grounds that it exemplifies IBE. He says that "nonrealists do not regard apparent explanatory success as sufficient reason to conclude truth, owing to underdetermination, etc., yet the inference to truth as the best explanation for the success of science makes a similar move at the metalevel" (Nickles, 2017, p. 159). My reply is that the psychological hypothesis that he invokes to discredit realism is a product of IBE, like other scientific hypotheses. Recall that it was designed to explain the conflicting reports of the two groups of the psychological subjects. Therefore, it is incoherent for Nickles to reject IBE and at the same time appeal to the psychological hypothesis (Park, 2019c, p. 64).

## 1.8 Ad Hoc

Frost-Arnold (2010) objects that "realism neither makes new predictions nor unifies previously disparate claims" (Frost-Arnold, 2010, p. 47). It is an ad hoc hypothesis similar to the famous ad hoc hypothesis in the history of science that opium has a dormative virtue. Just as this ad hoc hypothesis explains why people sleep after taking opium, but no other phenomena, so realism explains why science is successful, but no other phenomena. The two hypotheses would be virtuous if they explained other phenomena, phenomena that they were not designed to explain.

Frost-Arnold's foregoing objection to realism is in line with other philosophers' objections to it. Mizrahi objects that realism fails "to yield independently testable predictions that alternative explanations for success do not yield" (Mizrahi, 2012, p. 137). Nickles also accuses realism of failing to make "a scientific prediction" (Nickles, 2017, p. 157). All of these philosophers espouse the view called "naturalism." It holds that there is no strict distinction between philosophical and scientific methodologies. Scientists set up hypotheses about the world and test them against the world. Analogously, philosophers set up hypotheses about science and test them against science. On the naturalist account, realism should exhibit the virtues that successful scientific theories exhibit.

This naturalist objection to realism is built upon two assumptions. The first assumption is that realism is not the epistemic thesis that T is warranted but rather the sematic thesis that T is true (Park, 2019a, p. 280). After all, it is obviously unreasonable to demand that epistemic realism should have independently testable predictions and should make scientific predictions. By contrast, it is reasonable to demand that semantic realism should have such virtues, just like a scientific hypothesis.

The second assumption is that Psillos's interpretation of the NMA is correct and that my interpretation of it is incorrect. Suppose that the NMA is distinct from scientists' arguments, and realism is a meta-hypothesis about successful theories, as Psillos claims. Under this interpretation, it makes sense to request that realism should unify disparate scientific claims, just as Isaac Newton's mechanics unifies disparate phenomena. By contrast, suppose that the NMA is reducible to scientists' arguments, and that realism is reducible to the assertion that plate tectonics, the theory of evolution, the kinetic theory, and the like are true. Under this reductive view, realism unifies disparate phenomena because successful theories unify disparate phenomena, pace Frost-Arnold (Park, 2017d, p. 61).

Even if realists grant that Psillos's interpretation of the NMA is correct, they can overcome Frost-Arnold's objection. As Wray (2007, p. 83) observes, the phrase "the success of science" has two meanings. The first one is that some theories in science are successful, and the second one is that current scientists have such theories. Realism was originally designed to explain the former (Putnam, 1975). However, it can also explain the latter. According to realism, scientists have successful theories, but not unsuccessful ones, "because they have retained true theories and discarded false theories" (Park, 2017d, p. 63). On this account, if scientists keep true theories and throw out false ones, they end up possessing successful theories, but not unsuccessful ones. It follows that realism unifies the two disparate explananda, implying that truth underlies them. Therefore, realism is not an ad hoc hypothesis.

## 1.9 Empiricism

Just as it is useful to distinguish between semantic and epistemic realism, so it is useful to distinguish between semantic and epistemic empiricism (Park, 2019a, p. 286). The former holds that T is EA. The latter holds that we can justifiably believe that T is EA. These two positions are distinct from constructive empiricism (van Fraassen, 1980), which is analyzed and criticized in Sect. 3.10 of Chap. 3. Just as I mean either semantic or epistemic realism by "realism," so I mean either semantic or epistemic empiricism by "empiricism" throughout this book. Keep in mind that "empiricism" does not refer to constructive empiricism in this book.

Empiricists might suggest that science is successful because some theories are EA (Musgrave, 1988, p. 242; Ladyman, 1999, p. 186). Thus, empiricists invoke empirical adequacy, while realists invoke truth, to explain success. Empiricism is better than realism from the perspective of economy. If you loathe forming false beliefs, you would prefer the former over the latter.

How should we evaluate the empiricist explanation? Frost-Arnold's objection to the realist explanation in Sect. 1.8 applies no less to it. The empiricist hypothesis that successful theories are EA is ad hoc because it does not explain other phenomena. In response, empiricists might say that the empiricist hypothesis was originally proposed to explain the success of some theories, but it unexpectedly explains their existence in current science. Scientists have successful theories, but not unsuccessful ones, because they accepted EA theories and rejected empirically inadequate ones. On this account, if scientists keep EA theories and throw out empirically inadequate ones, they end up possessing successful theories, but not unsuccessful ones (Park, 2017d, p. 65). In sum, empiricists can get around Frost-Arnold's possible objection to them in the way realists would.

However, there is a conceptual problem with the empiricist explanation. On Laudan's (1981, p. 23) definition of success cited in Sect. 1.2, the success of T entails that *some* of its testable implications are true. By contrast, to say that it is EA means that *all* of them are true. Accordingly, the empiricist explanation can be compared to the explanation that "some crows are black because all crows are" (Musgrave, 1988, p. 242). What is wrong with such an explanation? Jarrett Leplin answers that "Generalizations do not explain their instances" (Leplin, 1997, p. 23). Ladyman retorts that Leplin's contention "needs to be argued rather than merely stated" (Ladyman, 1999, p. 186).

Leplin's assertion can be argued as follows (Park, 2014b, pp. 285–286). Suppose that there are two crows, $c_1$ and $c_2$, in the world. Why is $c_1$ black? Empiricists suggest that $c_1$ is black because $c_1$ and $c_2$ are black, which amounts to the suggestion that $c_1$ is black because $c_2$ is black. The problem with this explanation is that the color of $c_2$ calls for explanation, given that it is similar to that of $c_1$. Why is $c_2$ black? Empiricists would suggest that $c_2$ is black because $c_1$ and $c_2$ are black, which amounts to the suggestion that $c_2$ is black because $c_1$ is black. However, it is circular to say that $c_1$ is black because $c_2$ is black and that $c_2$ is black because $c_1$ is black. Such an explanation cannot dispel the original puzzlement over why $c_1$ is black.

## 1.10 Surrealism

Antirealists might contend that observables behave as if T were true. Leplin (1987) calls this view "surrealism" (a surrogate for realism). Surrealists claim that T is successful because phenomena are as if it were true. This explanation entails not that T is true but rather that it is EA (Musgrave, 1988, p. 243; Stanford, 2000, p. 268). Hence, surrealists can meet the demand to explain the success of T while disbelieving that it is true. Many renowned philosophers take surrealism to be a serious antirealist alternative to realism. They are Fine (1986, p. 160), Leplin (1987), Musgrave (1988, p. 243, 2017, p. 76), André Kukla (1998, p. 22), Stanford (2000, p. 268), Mizrahi (2012, p. 133), and Lyons (2003, p. 897, 2018, p. 147). For example, Fine says that unlike realists, antirealists treat "theories as if they were true, and their entities as if they were real" (Fine, 1986, p. 160).

Let me compare surrealism with empiricism as discussed in Sect. 1.9. They are similar in that both assert that T is EA, but not that T is true. However, they invoke distinct explanantia to explain the success of T. Surrealism invokes observables, whereas empiricism invokes empirical adequacy, which is a semantic property. Suppose that you are puzzled over why T is successful. Surrealism suggests that you should investigate the world, whereas empiricism suggests that you should investigate T, to dispel your puzzlement (Park, 2016b, p. 606). In short, surrealism and empiricism assert that the explanans for the success of T lies in the world and in T, respectively.

In Sect. 1.9, I argued that the empiricist explanation is circular. Recall that there is no significant difference between the explanandum and the explanans of the empiricist explanation. Accordingly, to explain the explanandum in terms of the explanans is to replace a mystery with a similar mystery. This criticism does not apply to the surrealist explanation, given that the success of T and the behavior of observables are substantially different from each other (Park, 2016b, p. 606). Of course, it is a mystery why observables behave as they do. But this new mystery is different in kind from the original mystery of why T is successful. Therefore, it is not vacuous to explain the old mystery in terms of the new mystery.

However, the surrealist explanation has its own problems. Let me use an analogy to expose them. Imagine that the Wright brothers make an airplane that flies after many trials and errors. What enables the airplane to fly? How can we explain the impressive performance of the airplane? The following two answers are possible. The first one is that the shape of the airplane is right. This explanation gives the credit to the Wright brothers, given that they designed the shape. The second answer is that air exists in the sky. This explanation gives the credit not to the Wright brothers but rather to the world. Such an explanation would disappoint the Wright brothers because it fails "to recognize their creativity and hard work" (Park, 2016b, p. 612).

Similarly, suppose that scientists work hard and devise T, and that it makes true predictions. Realists claim that it makes true predictions because it is true. This explanation acknowledges scientists' originality, given that they designed it. By contrast, surrealists claim that it is successful due to the way observables behave. This explanation gives the credit not to scientists but rather to observables. It fails to recognize scientists' originality (Park, 2016b, pp. 611–613). Scientists would not be happy with it any more than the Wright brothers would be happy with the explanation that their airplane could fly due to the existence of air.

Imagine that surrealists give the surrealist explanation of why T is successful. How would authors of T, viz., scientists who have devised T, respond to surrealists? Scientists would say to surrealists, "Surrealism is successful not because it is true, but rather because science behaves as if surrealism were true. If science did not behave as if surrealism were true, surrealism would not be successful. Therefore, scientists deserve credit for the success of surrealism." Surrealists might be unhappy with this explanation because it overlooks their originality and hard work. If they are, however, they ought not to give the surrealist explanation of why T is successful.

Let me move to Lyons's variant of surrealism. He asserts that a theory is successful because the "mechanisms postulated by the theory and its auxiliaries would, if actual, bring about all relevant phenomena thus far observed and some yet to be observed at time t" (Lyons, 2003, p. 900). This explanation appeals to the mechanisms of a theory, and thus it appears to be a realist explanation. On close examination, however, it is not a realist explanation because it does not assert that the mechanisms are actual. It only asserts that if they were actual, they would bring about certain phenomena. Lyons (2018, p. 147) adheres to this form of surrealism, taking it as a viable alternative to realism.

Is Lyons's variant tenable? It is not explanatory at all, although it appears to be explanatory. To put it differently, it does not give rise to the understanding of why a theory is successful, although it appears that it does. Let me use an analogy to expose a conceptual flaw within it. Consider the following two explanations:

(H1) Jane is elated because she got a job, and if she got a job, she would be elated.
(H2) Jane is elated because if she got a job, she would be elated.

(H1) gives rise to the understanding of why an explanandum occurs, whereas (H2) does not. In (H1), the two explanantia provide a reason for thinking that the explanandum occurs, whereas in (H2), the explanans does not provide any reason for thinking that the explanandum occurs. As Carl Hempel (1966, p. 48) argues, an explanans should provide good grounds for believing that an explanandum occurs. (H1) meets this requirement, whereas (H2) does not. Consequently, (H1) is explanatory, whereas (H2) is not. The same holds for (S1) and (S2):

(S1) T is successful because its mechanisms are actual, and if they were actual, they would bring about certain phenomena.
(S2) T is successful because if its mechanisms were actual, they would bring about certain phenomena.

In (S1), the two explanantia provide a reason for thinking that the explanandum occurred. However, it is not available for antirealists because it appeals to the realist hypothesis that the mechanisms are real. In (S2), the explanans does not provide a reason for thinking that the explanandum occurs. Therefore, (S2) is not explanatory

(Park, 2020, pp. 67–69). Other problems with Lyons's variant of surrealism are explicated in Seungbae Park (2020, pp. 67–69).

Surrealism seems to overcome Leplin's objection to empiricism. What is Leplin's objection to empiricism? Empiricism claims that T is successful because it is EA. Leplin objects that "empirical adequacy is an attribute that itself cries out for explanation" (Leplin, 1997, p. 23). The idea is that if success calls for explanation, empirical adequacy also calls for explanation. Empiricists, however, cannot invoke truth to explain empirical adequacy because they would become realists the moment they do so. Only realists can say that T is EA because it is true. In this context, surrealists might suggest that T is EA because phenomena are as if it were true. It appears, therefore, that surrealists can explain empirical adequacy without invoking truth.

What are we to make of the surrealist explanation that T is EA because observable parts of the world operate as if T were true? Evaluating it requires that we carefully compare what realists and surrealists do when they give the following explanations, respectively:

(R) T is EA because T is true.
(S) T is EA because observable parts of the world operate as if it were true.

Realists and surrealists are engaged in different sorts of businesses when they give (R) and (S), respectively. Realists are *explaining* why T is EA, whereas surrealists are *clarifying* what it is for T to be EA. Let me flesh out this difference in detail.

When realists give (R), they are not analyzing the concept of empirical adequacy, i.e., they are not making explicit what it is for T to be EA. They are rather *explaining* why T is EA. The explanation involves an ampliative inference, the inference from the premise that T is EA to the conclusion that it is true. This inference is inductive, given that EA theories might be false. It follows that an epistemic risk is involved in (R). Antirealists, who are reluctant to run the epistemic risk, would reject (R).

By contrast, when surrealists give (S), they are not making any ampliative inference. They are rather *analyzing* the concept of empirical adequacy, making explicit how empirical adequacy relates to the behavior of observables. They are claiming that T is EA due to the behavioral pattern of observables, implying that if the behavioral pattern were different, T would not be EA. Relatedly, if you are already familiar with the concept of empirical adequacy, you would not raise any objection to (S). You would not say that even if T is EA, observables might not behave as if it were true. If you say so, you would only betray your ignorance about the relationship between empirical adequacy and the behavioral pattern of observables.

What can we conclude from the fact that realists and surrealists are engaged in different sorts of businesses when they give (R) and (S)? (S) cannot be an alternative to (R). To provide (S) as an alternative to (R) is just as inappropriate as to provide an orange in the context where different varieties of apples compete with each other (Park, 2019d, pp. 386–390). Relatedly, (S) might be useful to philosophy professors who aim to convey the concept of empirical adequacy to their students. However, (S) would not be useful to antirealists who aim to undermine (R) by proposing an antirealist alternative to (R). Accordingly, (S) cannot be an adequate response to

Leplin's objection to empiricism that "empirical adequacy is an attribute that itself cries out for explanation" (Leplin, 1997, p. 23).

## 1.11 Predictive Similarity

Stanford puts forward an original proposal, according to which "the success of a given false theory in a particular domain is explained by the fact that its predictions are (sufficiently) close to those made by the true theoretical account of the relevant domain" (Stanford, 2000, p. 275). For example, the caloric theory was successful due to its predictive similarity to the kinetic theory. Both the caloric and kinetic theories predicted that if a hot piece of iron and a cold piece of iron were in contact with each other, the former would become less hot and the latter would become less cold. It was due to such predictive similarities that the caloric theory was successful. Mizrahi (2020, p. 140) appeals to Stanford's account to support his original position called "relative realism."

Stanford's account has a few problems. It competes with an alternative that the success of T is due to its predictive similarity to a false theory. Consider three rival theories, $T_1$, $T_2$, and $T_3$, where $T_1$ and $T_2$ are false and $T_3$ is true. Given that they all make similar predictions, they are all predictively similar to one another. On Stanford's account, the success of $T_1$ is due to its predictive similarity to $T_3$. However, it competes with the alternative explanation that the success of $T_1$ is due to its predictive similarity to $T_2$. Stanford faces this question: why is appealing to a true theory better than appealing to a false theory?

Let me turn to the problem of circularity. Consider again $T_1$, $T_2$, and $T_3$ from above. Why is $T_3$ successful? Since $T_3$ is a true theory, Stanford would have to suggest that its success is due to its predictive similarity to itself. Such an explanation is vacuous, casting no light on why $T_3$ is successful (Park, 2003, p. 165). In light of this criticism, Stanford might suggest that the success of $T_3$ is due to its predictive similarity to $T_1$. However, critics would challenge him to explain why $T_1$ is successful. He might suggest that the success of $T_1$ is due to its predictive similarity to $T_2$. Critics would now ask why $T_2$ is successful. Stanford cannot answer that the success of $T_2$ is due to its predictive similarity to $T_3$ because such an answer amounts to saying that the success of $T_3$ is due to its predictive similarity to itself.

Let me turn to another problem with Stanford's proposal. Consider again $T_1$, $T_2$, and $T_3$ from above. Suppose $T_1$ is false and conceived whereas $T_3$ is true and unconceived. Why is $T_1$ successful? Stanford says that the success of $T_1$ is due to its predictive similarity to $T_3$. In my view, however, it is inadequate to invoke $T_3$ to explain the success of $T_1$. $T_3$ is not even conceived, and thus the success of $T_1$ is independent of $T_3$. Even if $T_3$ is conceived, the success of $T_1$ is independent of $T_3$. Admittedly, we can infer the success of $T_1$ from its predictive similarity to $T_3$. It does not follow, however, that the predictive similarity explains the success of $T_1$. Inference and explanation are different affairs, as Sylvain Bromberger's (1966) flagpole objection to Hempel's (1966) deductive-nomological model shows.

Readers who are not yet convinced of this objection to Stanford's proposal are encouraged to consider the explanation that Mars is round because it is similar to Venus vis-à-vis its shape. This explanation is not adequate because Mars can be round independently of whether or not Venus is round. Analogously, consider the explanation that $T_1$ is successful because it is similar to $T_3$ vis-à-vis predictions. This explanation is not adequate because $T_1$ is successful independently of whether or not $T_3$ is successful (Park, 2003, p. 166).

## 1.12 Summary

Semantic realism is explicit and epistemic realism is implicit in the NMA. Laudan and Lyons think that past theories are counterexamples that refute realism. However, the putative counterexamples are compatible with semantic and epistemic realism. Antirealists might reject the demand to explain the success of theories and hence the NMA. To do so, however, violates the basic rule for attacking an argument. Van Fraassen, Laudan, and Fine reject IBE and hence the NMA, but their rejection of IBE backfires on their own positive explanations. It is self-defeating for Wray and van Fraassen to offer an evolutionary alternative to the realist explanation. It is also self-refuting for Nickles to appeal to a psychological hypothesis to discredit realism. Realism explains why some theories make true predictions and why current scientists have such theories. Consequently, realism is not an ad hoc hypothesis, pace Frost-Arnold. It is vacuous for empiricists to invoke empirical adequacy to explain success. The surrealist explanation fails to recognize scientists' originality. Predictive similarity is an inadequate explanans for success. In short, the prominent objections to the NMA are problematic.

## References


Akins, K. (1996). Of sensory systems and the 'aboutness' of mental states. *The Journal of Philosophy, 93*(3), 337–372.

Bromberger, S. (1966). Why-questions. In R. Colodny (Ed.), *Mind and cosmos: Essays in contemporary science and philosophy* (pp. 86–111). University of Pittsburgh Press.

Darwin, C. (1859/1993). *The portable Darwin*. In P. Graham (Ed.), Duncan Porter and Paris: Penguin Books.

Fine, A. (1986). Unnatural attitudes: Realist and instrumentalist attachments to science. *Mind, 95*(378), 149–179.

Fine, A. (1991). Piecemeal realism. *Philosophical Studies, 61*(1–2), 79–96.

Frost-Arnold, G. (2010). The no-miracles argument: Inference to an unacceptable explanation. *Philosophy of Science, 77*(1), 35–58.

Goldman, A., & Beddor, B. (2016). Reliabilist epistemology. In E. N. Zalta (Ed.), *The Stanford encyclopedia of philosophy*. https://plato.stanford.edu/archives/win2016/entries/reliabilism/

Hempel, C. (1966). *Philosophy of natural science*. Prentice-Hall.

Hendry, R. (2001). Are realism and instrumentalism methodologically indifferent? *Philosophy of Science, 68*(Proceedings), S25–S37.
Hume, D. (1888/1978). *A treatise of human nature*. L. A. Selby-Bigge and P. H. Nidditch (Eds.), : Oxford University Press.
Khalifa, K. (2010). Default privilege and bad lots: Underconsideration and explanatory inference. *International Studies in the Philosophy of Science, 24*(1), 91–105.
Kukla, A. (1998). *Studies in scientific realism*. Oxford University Press.
Ladyman, J. (1999). Review. A novel defense of scientific realism. J. Leplin. *The British Journal for the Philosophy of Science, 50*(1), 181–188.
Ladyman, J., Douven, I., Horsten, L., & van Fraassen, B. (1997). A defense of van Fraassen's critique of abductive inference: Reply to Psillos. *The Philosophical Quarterly, 47*(188), 305–321.
Laudan, L. (1981). A confutation of convergent realism. *Philosophy of Science, 48*(1), 19–49.
Laudan, L. (1984). Explaining the success of science: Beyond epistemic realism and relativism. In J. Cushing, C. Delaney, & G. Gutting (Eds.), *Science and reality* (pp. 83–105). University of Notre Dame Press.
Leplin, J. (1987). Surrealism. *Mind, 97*(384), 519–524.
Leplin, J. (1997). *A novel defense of scientific realism*. Oxford University Press.
Lyons, T. (2003). Explaining the success of a scientific theory. *Philosophy of Science, 70*(5), 891–901.
Lyons, T. (2005). Toward a purely axiological scientific realism. *Erkenntnis, 63*(2), 167–204.
Lyons, T. (2016). Scientific realism. In P. Humphreys (Ed.), *The Oxford handbook of philosophy of science* (pp. 564–584). Oxford University Press.
Lyons, T. (2017). Epistemic selectivity, historical threats, and the non-epistemic tenets of scientific realism. *Synthese, 194*(9), 3203–3219.
Lyons, T. (2018). Four challenges to epistemic scientific realism. *Spontaneous Generations: A Journal for the History and Philosophy of Science, 9*(1), 146–150.
Mizrahi, M. (2012). Why the ultimate argument for scientific realism ultimately fails. *Studies in History and Philosophy of Science Part A, 43*(1), 132–138.
Mizrahi, M. (2013). The pessimistic induction: A bad argument gone too far. *Synthese, 190*(15), 3209–3226.
Mizrahi, M. (2020). *The relativity of theory: Key positions and arguments in the contemporary scientific realism/antirealism debate*. Springer.
Musgrave, A. (1988). The ultimate argument for scientific realism. In R. Nola (Ed.), *Relativism and realism in science* (pp. 229–252). Kluwer Academic Publishers.
Musgrave, A. (2017). Strict empiricism versus explanation in science. In E. Agazzi (Ed.), *Varieties of scientific realism: Objectivity and truth in science* (pp. 71–93). Springer.
Nickles, T. (2017). Cognitive illusions and nonrealism: Objections and replies. In E. Agazzi (Ed.), *Varieties of scientific realism: Objectivity and truth in science* (pp. 151–163). Springer.
Park, S. (2001). *Scientific realism vs. scientific antirealism*. PhD dissertation. University of Arizona.
Park, S. (2003). Ontological order in scientific explanation. *Philosophical Papers, 32*(2), 157–170.
Park, S. (2011). Coherence of our best scientific theories. *Foundations of Science, 16*(1), 21–30.
Park, S. (2014a). On the evolutionary defense of scientific antirealism. *Axiomathes, 24*(2), 263–273.
Park, S. (2014b). The doxastic requirement of scientific explanation and understanding. *Prolegomena, 13*(2), 279–290.
Park, S. (2015a). Against motivational efficacy of beliefs. *Coactivity: Philosophy Communication, 23*(1), 86–95.
Park, S. (2015b). Explanatory failures of relative realism. *Epistemologia, 38*(1), 16–28.
Park, S. (2016a). Extensional scientific realism vs. intensional scientific realism. *Studies in History and Philosophy of Science, 59*, 46–52.
Park, S. (2016b). Realism versus surrealism. *Foundations of Science, 21*(4), 603–614.
Park, S. (2017a). Scientific antirealists have set fire to their own houses. *Prolegomena, 16*(1), 23–37.
Park, S. (2017b). Defense of epistemic reciprocalism. *Filozofija. Sociologija, 28*(1), 56–64.

Park, S. (2017c). Problems with using evolutionary theory in philosophy. *Axiomathes, 27*(3), 321–332.
Park, S. (2017d). The unificatory power of scientific realism. *Disputatio, 9*(44), 59–73.
Park, S. (2018). Justifying the special theory of relativity with unconceived methods. *Axiomathes, 28*(1), 53–62.
Park, S. (2019a). The descriptive and normative versions of scientific realism and pessimism. *Filozofia: Journal for Philosophy, 74*(4), 278–290.
Park, S. (2019b). In defense of realism and selectivism from Lyons's objections. *Foundations of Science, 24*(4), 605–615.
Park, S. (2019c). Scientific realism and the future development of science. *Diametros: An Online Journal of Philosophy, 60*, 61–71.
Park, S. (2019d). Surrealism is not an alternative to scientific realism. *Logos & Episteme: An International Journal of Epistemology, 10*(4), 379–393.
Park, S. (2020). Critiques of axiological realism and surrealism. *Acta Analytica, 35*(1), 61–74.
Psillos, S. (1996). Scientific realism and the 'pessimistic induction.' *Philosophy of Science, 63*(Proceedings), S306–S314.
Psillos, S. (2011). Making contact with molecules: On Perrin and Achinstein. In G. Morgan (Ed.), *Philosophy of science matters: The philosophy of Peter Achinstein* (pp. 177–190). Oxford University Press.
Putnam, H. (1975). *Mathematics, matter and method: Philosophical papers, volume 1*. Cambridge University Press.
Quoidbach, J., Gilbert, D., & Wilson, T. (2013). The end of history illusion. *Science, 339*(6115), 96–98.
Sankey, H. (2017). Realism, progress and the historical turn. *Foundations of Science, 22*(1), 201–214.
Stanford, K. (2000). An antirealist explanation of the success science. *Philosophy of Science, 67*(2), 266–284.
van Fraassen, B. (1980). *The scientific image*. Oxford University Press.
van Fraassen, B. (1989). *Laws and symmetry*. Oxford University Press.
Wray, B. (2007). A selectionist explanation for the success *and failures* of science. *Erkenntnis, 67*(1), 81–89.
Wray, B. (2008). The argument from under consideration as grounds for anti-realism: A defence. *International Studies in the Philosophy of Science, 22*(3), 317–326.
Wray, B. (2010). Selection and predictive success. *Erkenntnis, 72*(3), 365–377.
Wray, B. (2012). Epistemic privilege and the success of science. *Noûs, 46*(3), 375–385.
Wray, B. (2013). The pessimistic induction and the exponential growth of science reassessed. *Synthese, 190*(18), 4321–4330.
Wray, B. (2015). Pessimistic inductions: Four varieties. *International Studies in the Philosophy of Science, 29*(1), 61–73.

# Chapter 2
# Critiques of the Pessimistic Induction

**Abstract** The pessimistic induction (PI) should not be conflated with Laudan's objection that there are historical counterexamples to realism, nor with the Humean argument that earlier theories were false, so current theories are unwarranted. It is arbitrary for antirealists to accept the PI but to reject inference to the best explanation. It is incoherent for antirealists to use evolutionary theory to bolster the PI. The PI has four intrinsic problems, on the basis of which I construct a second-order PI that I call "the grand pessimistic induction," which spells doom for the first-order PIs that are directed at scientists, realists, and scientific theories.

**Keywords** Evolutionary theory · Humean argument · Inference to the best explanation · Misleading evidence · Pessimistic induction · Proportional pessimism · Unconceived alternatives

## 2.1 Introduction

The pessimistic induction (PI) is regarded as the most forceful argument for antirealism, as we noted in Chap. 1. This chapter aims to answer the following questions about the PI: What exactly does the PI say? Can we accept the PI but reject inference to the best explanation (IBE)? Can we use evolutionary theory to reinforce the PI? What intrinsic problems does the PI have? What variants of the PI are there in the literature? Are they cogent? My answers to these questions will all be disagreeable to antirealists.

The structure of this chapter is as follows. In Sect. 2.2, I analyze the PI, arguing that it should not be confounded with Larry Laudan's (1981) objection that was discussed in Sect. 1.3 of Chap. 1, nor with what I call "the Humean argument," according to which since earlier theories were false, current theories are unwarranted. In Sect. 2.3, I argue that it is illegitimate for antirealists to accept the PI and to reject IBE and that it is self-defeating for antirealists to use evolutionary theory to augment the PI. In Sect. 2.4, I spell out four intrinsic problems with the PI and then construct a second-order PI that I call "the grand pessimistic induction" (the


S. Park, *Embracing Scientific Realism*, Synthese Library 445, https://doi.org/10.1007/978-3-030-87813-9_2

grand PI). The grand PI has disastrous implications for the first-order PIs that are directed at scientists, realists, and scientific theories.

## 2.2 The Pessimistic Induction

The PI has several variants. Let me call the oldest one "the classic PI." It is directed at scientific theories, and it is advocated not only by some philosophers but also by some scientists. For example, Henri Poincaré states the following:

> The ephemeral nature of scientific theories takes by surprise the man of the world. Their brief period of prosperity ended, he sees them abandoned one after the other; he sees ruins piled upon ruins; he predicts that the theories in fashion today will in a short time succumb in their turn, and he concludes that they are absolutely in vain. (Poincaré, 1905/1952, p. 160)

Similarly, Hilary Putnam says that "just as no term used in the science of more than fifty (or whatever) years ago referred, so it will turn out that no term used now (except maybe observational terms, if there are such) refers" (Putnam, 1978, p. 25). Pessimists appeal to lists of theories from the past, such as Laudan's (1981, p. 33) and Kyle Stanford's (2006, pp. 19–20) lists, to establish the premise of the classic PI that past theories were false. From this premise, they draw the conclusion that it is likely[1] that current theories are also false.

The classic PI should not be conflated with Laudan's objection (Laudan, 1981; Lyons, 2017, 2018), which was discussed in Sect. 1.3 of Chap. 1. The classic PI claims that most current theories are false, while Laudan's objection states that some historical counterexamples rebut the universal statement that all successful theories are true. Laudan's objection is compatible with realism, while the classic PI is incompatible with it. For this reason, antirealists should not be content merely with Laudan's objection, but should aim for the classic PI (Park, 2019a, p. 608).

Laudan (1981) does not run "the pessimistic induction against current theories" (Park, 2011a, p. 22). Many writers, however, continue to falsely attribute the classic PI to Laudan (1981). They include James Ladyman (1999, p. 186), Peter Vickers (2017, p. 3221), Jamin Asay (2019, p. 591), and Finnur Dellsén (2019, p. 32). Ladyman writes, for instance, "The pessimistic meta-induction from the history of past theory change in science was originally proposed to confute the realist's claim that truth explains the success of science by showing that our best present and future theories are most likely false (Laudan [1981])" (Ladyman, 1999, p. 186). All of these writers are wrong to attribute the classic PI to Laudan (1981). I am certain that Laudan (1981) does not even construct an inductive argument, to say nothing of the classic PI.

We should separate the classic PI not only from Laudan's objection but also from the Humean argument. The Humean argument holds that since outdated theories were *false*, current theories are *unwarranted*. This argument is similar to the

[1] I drop "it is likely" hereafter for the sake of brevity.

argument that since stones have fallen down, we should be skeptical about whether they will fall down in the future. By contrast, the classic PI is similar to the inductive argument that since stones have fallen down, they will fall down in the future. The classic PI accepts, while the Humean argument rejects, David Hume's (1888/1978, p. 79) uniformity principle that the future is similar to the past. Therefore, the classic PI and the Humean argument conflict with each other (Park, 2019b, pp. 142–144).

How do the Humean argument and Laudan's objection relate to each other? Laudan's objection states that some historical counterexamples refute the universal statement that all successful theories are true, whereas the Humean argument states that since earlier theories were false, current theories are unwarranted. A similarity between them is that they are both based upon the same historical fact that some theories were false. A difference between them is that Laudan's objection is reticent about whether current theories are unwarranted, whereas the Humean argument states that they are unwarranted. Another difference between them is that Laudan's objection is compatible with realism, whereas the Humean argument is incompatible with it.

Why do I fuss about the distinction between the classic PI and the Humean argument? Antirealists tend to retreat from the classic PI to the Humean argument when the classic PI is under fire (Park, 2019b, p. 143). For example, realists point out that theories in current science are *more successful* than those in past science, so it is illegitimate to expect the downfall of the former based on that of the latter (Musgrave, 1985, p. 211; Leplin, 1997, p. 141; Doppelt, 2007, p. 111, 2014; Devitt, 2011; Fahrbach, 2011a, 2011b; Park, 2011b; Mizrahi, 2013). Antirealists acknowledge this difference between current theories and their predecessors. However, they say that the falsity of outdated theories does demonstrate that current theories are unwarranted. On their account, current theories might be true, but it does not matter whether they are true. Even if they are true, they are unwarranted. Past science, which is the graveyard of discarded theories, makes current theories unwarranted.

Antirealists' retreat from the classic PI to the Humean argument has two problems. First, some antirealists provide their own accounts for the success of science, as we have seen in Chap. 1. For example, surrealists claim that theories are successful because observables operate as if they were true. When they give such an account, they are making an ampliative inference from the success of theories to the behavior of observables. Their inference conflicts with the Humean argument. Humean skeptics would not infer that observables operate as if successful theories were true from the observation that theories are successful. Therefore, the retreat from the classic PI to the Humean argument is not available to those who give antirealist explanations of the success of science.

Second, some antirealists embrace instrumentalism. Stanford, for example, asserts that "we might use our theories for prediction, intervention, and other pragmatic purposes without believing the theoretical descriptions they offer of the natural world" (Stanford, 2006, p. 197). However, instrumentalism clashes with the Humean argument. Instrumentalists believe that a theory that was useful will continue to be useful, whereas Humean skeptics are skeptical about whether it will continue to be useful. Humean skeptics reject instrumentalists' ampliative inference, pointing out that it relies on the uniformity principle. Therefore, instrumentalists cannot retreat from the classic PI to the Humean argument (Park, 2019b, p. 143).

## 2.3 Problematic Positions

### *2.3.1 The Classic PI and IBE*

Some pessimists accept the classic PI but reject IBE. In this subsection, I argue that it is arbitrary for them to do so.

Laudan writes, "Most of the past theories of science are already suspected of being false; there is presumably every reason to anticipate that current theories of science will suffer a similar fate" (Laudan, 1977, p. 126). However, he rejects IBE and accuses the no-miracles argument of being "a monumental case of begging the question" (Laudan, 1981, p. 45). Brad Wray endorses the classic PI, saying "that future developments in a field will reveal additional competing theories that are also able to account for the data, theories, though, that are now *unconceived* alternatives" (Wray, 2010, p. 371). However, Wray (2008, 2012) vigorously defends the argument from underconsideration, which Bas van Fraassen (1989, p. 143) originally used to attack IBE. Thomas Nickles also embraces the classic PI, claiming that scientific theories can change radically with the passage of enough time. He says that "over a long enough time the changes can be radical" (Nickles, 2017, p. 159). However, he rejects IBE, saying that "nonrealists do not regard apparent explanatory success as sufficient reason to conclude truth, owing to underdetermination, etc., yet the inference to truth as the best explanation for the success of science makes a similar move at the metalevel" (Nickles, 2017, p. 159).

It is arbitrary for these philosophers to accept the classic PI but to reject IBE. The classic PI exemplifies a rule of inference called "inductive generalization." Inductive generalization makes an inference from observed instances to unobserved instances. Pessimists observe the behavior of past theories and then predict the behavior of current theories. Laudan and Nickles accept inductive generalization, but they reject IBE. Consequently, they have the burden of explaining why they take different attitudes toward inductive generalization and IBE. It is not clear how they would respond to their possible opponents, those who claim that IBE is reliable, but that inductive generalization is not.

Laudan, Wray, and Nickles might be tempted to argue that there are unconceived alternatives to a hypothesis, so IBE is not reliable, whereas there is no unconceived alternative to a general statement, so inductive generalization is reliable. However, this reply is indefensible. Suppose that you have observed ten copper bars conducting heat, and you conclude that all copper bars conduct heat. Your conclusion competes with an alternative that all copper bars conduct heat except for one copper bar, another alternative that all copper bars conduct heat except for two copper bars, and so on. For this reason, Stathis Psillos (1997, p. 370) claims that we take an epistemic risk when we make an inference from success of empirical adequacy.

### 2.3.2 *The Classic PI and Evolutionary Theory*

Some pessimists appeal to evolutionary theory to strengthen the classic PI. In this subsection, I argue that it is self-defeating to do so.

Thomas Kuhn (1962/1970) compares the development of science with the evolution of organisms. According to evolutionary theory, natural selection is a goal-free process. Natural selection does not have the goal, for example, of producing long-necked giraffes. Analogously, scientific development consists in the repeated alternations of normal and revolutionary science, and these repeated alternations are goal-free processes, i.e., they do not lead us to truths. The analogy between the evolution of organisms and science is "very nearly perfect" (Kuhn, 1962/1970, p. 172). Therefore, current theories are no more truthlike than their forerunners.

This antirealist tradition of using evolutionary theory to support the pessimistic accounts of scientific development continues. On the one hand, for example, Wray defends the pessimistic account of scientific development, according to which science loses old concepts and acquires new ones when a scientific revolution occurs (Wray, 2011, p. 136). On the other hand, he appeals to evolutionary theory, saying that just as organisms lose old morphologies and acquire new ones when speciation occurs, so science loses old concepts and acquires new ones when a scientific revolution occurs. He states that waste and "destruction are thus as prevalent in the biological world as they appear to be in the scientific world" (Wray, 2011, p. 136).

Nickles invokes the classic PI and evolutionary theory to refute selective realism (Worrall, 1989; Kitcher, 1993; Psillos, 1999), which emphasizes that new theories inherit some theoretical assumptions from old theories. (Selective realism is analyzed in detail in Sect. 5.6.1 of Chap. 5.) Evolutionary theory claims that small changes in organisms can be accumulated and that given enough time, future organisms will be radically different from present organisms. Nickles says that "it is a basic lesson from evolutionary biology that very slow evolution can be as transformative as you please, given enough time" (Nickles, 2017, p. 158). Similarly, even if some of the assumptions of current theories are preserved, future theories will be radically different from current theories, given enough time. He writes, "You can have all the continuity you want between temporally adjacent work, but over a long enough time, the changes can be radical" (Nickles, 2017, p. 159). This objection to selective realism appears to be a lethal one.

Yet, it is incoherent for Kuhn, Wray, and Nickles to use evolutionary theory to flesh out their pessimistic accounts of scientific development (Park, 2017a, pp. 324–326). Their pessimistic accounts imply that evolutionary theory is false and that it is no closer to the truth than its predecessors. Any philosophical account of scientific development that relies on a false scientific theory is false. Therefore, their philosophical accounts of scientific development are false.

How would Kuhn, Wray, and Nickles respond to this criticism? They might withdraw their appeal to evolutionary theory, but stick to their pessimistic accounts of scientific development. Specifically, Kuhn might discard the scientific view that organisms do not develop toward a goal, but retain the philosophical view that science does not develop toward truths. Analogously, Wray might reject the scientific view that organisms lose old morphologies and acquire new morphologies when speciation occurs, but retain the philosophical view that science loses old concepts and acquires new ones when a scientific revolution occurs. Similarly, Nickles might abandon the scientific claim that given enough time, organisms change radically, but he might insist that given enough time, scientific theories change radically.

A problem with these moves is that the scientific views are similar to their philosophical counterparts, so they should be treated equally. It is arbitrary to reject the scientific views but to accept the philosophical views. It is not clear what Kuhn, Wray, and Nickles could say to scientists who equally arbitrarily reject the philosophical views but accept the scientific views. For example, scientists might accept the scientific view that given enough time, organisms change radically, but reject the philosophical view that given enough time, scientific theories, including evolutionary theory, change radically. Blunt scientists would say to philosophers that it is arrogant and self-serving for philosophers to reject the scientific views but to accept the philosophical views.

## 2.4 The Grand PI

### *2.4.1 Four Intrinsic Problems*

Let me turn to four objections to the classic PI. First, as we have seen in Sect. 2.2 of this chapter, many philosophers argue that theories in current science are empirically superior to those in past science, so it is fallacious to infer the downfall of the former from that of the latter. This objection to the classic PI has an intuitive appeal. Ptolemaic scientists believed that Jupiter revolved around the earth, whereas Copernican scientists believed that Jupiter revolved around the sun. We have the Hubble Space Telescope, but Ptolemaic scientists did not. It is implausible that Copernican scientists' beliefs will be overturned just as Ptolemaic scientists' beliefs were overturned.

Second, those who make use of Laudan's (1981, p. 33) list to support the classic PI commit the fallacy of biased statistics (Fahrbach, 2011a, 2011b; Park, 2011b; Mizrahi, 2013, 2015, 2016). On the list, we find theories from the distant past, but not from the recent past. The former are theories that were abandoned before the twentieth century, whereas the latter are theories that have been retained since the early twentieth century. The number of the latter is much larger than that of the former, and thus most theories from the past have not been overthrown. Ludwig Fahrbach writes, "at least 95% of all scientific work ever done has been done since

1915" (Fahrbach, 2011a, p. 149). Similarly, I note that "the body of scientific knowledge exploded in the 20th century" (Park, 2011b, p. 79). Moti Mizrahi (2013, 2016) performs an experiment of randomly selecting sample theories from the general population of past theories. The outcome of his experiment is that most of the selected theories are not abandoned in the twenty-first century. These optimists' historical observations rebut the premise of the classic PI.

Third, scientific revolutions occur often in some fields of science, but not in others. For example, several theories of light were accepted and rejected from the time of Isaac Newton up to the twentieth century, but the Ptolemaic theory persisted for about 1200 years (Fahrbach, 2011a, pp. 141–142). However, the classic PI lumps together all outdated theories from different fields of science, asserting that since they were false, their successors are likewise all false (Lange, 2002). It is problematic "to indiscriminately group together theories developed under very different circumstances, and thus generalize from an unnatural grouping" (Wray, 2015, p. 64).

Fourth, the classic PI invoking Laudan's list overlooks the important change that occurred in science at around the turn of the twentieth century (Park, 2011b, pp. 80–83). During this period, many thick and high walls separating the different fields of science began to fall down. In other words, scientists of many fields of science began to take into account the research outcomes of neighboring fields of science when they set up hypotheses in their own fields of science. Geologists, for example, when theorizing about distant past events, began to use the radiometric dating method that physicists developed. In a nutshell, many fields of science began to put *neighbor constraints* on one another, and as a result, the birth qualities of hypotheses became better than the birth qualities of those that were proposed before this period.

In the next subsection, I utilize these four intrinsic flaws with the classic PI to construct a second-order PI against the first-order PIs that are directed at scientists, realists, and scientific theories, including Stanford's (2006) problem of unconceived alternatives (PUA).

### 2.4.2 The PUA

The PUA is a successor to the classic PI. It states that past scientists lacked the cognitive ability to think of alternatives to their theories, so current scientists must also lack the cognitive ability to think of alternatives to their theories. For example, germ theorists are unable to conceive of alternatives to the germ theory any more than miasma theorists were able to conceive of alternatives to the miasma theory. On this account, current theories, which were once unconceived alternatives to past theories, superseded past theories. As a result, future theories, which are unconceived alternatives to current theories, will supersede current theories.

The PUA does not make the Humean inference that because earlier scientists could not ideate alternatives to their theories, we do not know whether current scientists cannot ideate alternatives to their theories. The PUA rather makes the inductive inference that since earlier scientists could not ideate competitors to their theories, current scientists cannot ideate competitors to their theories either. It distorts the PUA to interpret it as merely asserting that we should be skeptical about current scientists' cognitive capacity to ideate alternatives to their theories. Keep in mind that the PUA and its predecessor, the classic PI, are built upon the uniformity principle.

The PUA differs from the classic PI in the following two respects. First, the classic PI concerns scientific theories, asserting that because past theories were thrown out, current theories will also be thrown out. By contrast, the PUA concerns scientists, claiming that since past scientists did not have the cognitive capacity to entertain alternatives to their theories, contemporary scientists also do not have the cognitive capacity to entertain alternatives to their theories. Second, the classic PI does not even invoke the notion of unconceived alternatives, to say nothing of whether there are finitely or infinitely many such competitors. By contrast, the PUA asserts that the possibility space of rivals "appears to be indeterminate and unbounded" (Stanford, 2006, p. 133). In other words, it appears that there are indefinitely or infinitely many unconceived rivals.

What motivates Stanford to allude to an infinity of alternatives? He aims to get around the first problem with the classic PI mentioned in Sect. 2.4.1 of this chapter, viz., the classic PI overlooks the fact that the evidence for current theories is stronger than the evidence for earlier theories. On his account, past and current scientists, due to their limited cognitive capacity, cannot exhaust the space of unconceived rival theories. They are all "creatures whose cognitive constitutions are not well suited to the task of exhausting the kinds of spaces of serious candidate theoretical explanations from which our scientific theories are drawn" (Stanford, 2006, p. 45). Consequently, current theories are fated to be overthrown, just as their precursors were, despite the fact that the former are empirically more justified than the latter.

How do I assess the PUA? Overall, it is an original and impressive argument. It is distinct from and improves upon the classic PI. This improvement echoes the improvements of scientific theories over their predecessors. Consequently, the PUA poses a greater threat to realism than does the classic PI, and realists should take the PUA seriously.

The PUA can be defused as follows. Scientists improve upon their predecessors, just as their theories do (Bird, 2007, p. 108; Park, 2011b, p. 83, 2017b, p. 617; Mizrahi, 2020, p. 98). Scientists learn from past mistakes, and they devise theories with the aim to avoid the pitfalls of their predecessors' theories. This cognitive difference between past and current scientists invalidates the inference that since past scientists lacked the cognitive ability to think of alternatives to their theories, current scientists must also lack the cognitive ability to think of alternatives to their theories.

The PUA can also be countered as follows. Recent past theories more or less coincide with current theories. The germ theory, for example, is a current theory, given that it is accepted in the early twenty-first century, but it is also a recent past theory, given that it was accepted in the early twentieth century. Germ theorists in the early twentieth century conceived of their future theory, which is the germ theory. This historical fact entitles us to construct the optimistic induction that since recent past scientists in the early twentieth century had the cognitive capacity to conceive of their future theories, current and future scientists also have the cognitive capacity to conceive of their future theories (Park, 2017b, p. 617).

On second thought, however, these criticisms against the PUA are gratuitous. The PUA is subject to a second-order PI that I call "the grand PI" (Park, 2018). The grand PI consists of the premise that the classic PI had infinitely many hidden problems and the conclusion that its successors, which are infinitely many, have infinitely many hidden problems. We cannot conceive of the infinitely many successors, but they do exist. The grand PI predicts that pessimists will construct a successor to the PUA, just as Stanford constructed the PUA, a successor to the classic PI.

What is the justification for the premise of the grand PI that the classic PI has infinitely many hidden problems? As noted earlier in Sect. 2.4.1 of this chapter, it turned out that the classic PI has the following four intrinsic problems:

**The List of Problems with the Classic PI**

- The classic PI overlooks the empirical superiority of current theories over past ones.
- Its sample is biased.
- It lumps together disparate theories.
- It overlooks neighbor constraints.

These problems were hidden when the pioneering pessimists of the twentieth century propounded the classic PI. The hidden problems were brought out as a result of the debate between realists and pessimists. It follows that there are more such problems with the classic PI. They are as of now hidden, but they will be unhidden as realists and pessimists continue to debate over the classic PI. Hence, there are infinitely many such problems (Park, 2018, p. 12). This reasoning is similar to that of the PUA. The PUA claims that there were unconceived alternatives to earlier theories, so there are unconceived alternatives to present and future theories. Hence, the number of unconceived alternatives is infinitely large. This inference of the PUA and the inference of the grand PI rise or fall together.

Seungbae Park (2018, p. 12) deliberately refrains from exposing the intrinsic problems with the PUA because it is supererogatory to do so. To use an analogy, imagine that there is a succession of dominos and that the first domino falls over. Under this circumstance, we do not need to push the second domino because it will fall over even if we do nothing. If we want, however, we can push it as well. In this sense, it is supererogatory to push the second domino. Similarly, there is a succession of two pessimist arguments: the classic PI and the PUA. Since the first one is infinitely problematic, the second one is infinitely problematic. We do not need to expose intrinsic problems with the second one. If we want, however, we can expose intrinsic problems with it as well. In this sense, it is superfluous to construct a list of the intrinsic problems with the PUA.

This inference echoes the inferences of the classic PI and the PUA. According to the classic PI, once it is established that $T_1$, a past theory, was false, it is established inductively that $T_2$, the present successor to $T_1$, is also false. Adherents to the classic

PI do not directly examine scientists' arguments for $T_2$ to infer that $T_2$ is false. They merely point out that $T_1$ was false. According to the PUA, once it is established that scientists of $T_1$ could not think of alternatives to $T_1$, it is established inductively that scientists of $T_2$ cannot think of alternatives to $T_2$. Adherents to the PUA do not directly investigate current scientists' cognitive capacity to infer that they cannot think of alternatives to $T_2$. They merely point out that scientists of $T_1$ could not conceive of alternatives to $T_1$.

The PUA is not the only successor to the classic PI that pessimists have constructed to improve upon the classic PI. They also have constructed or can construct other successors to the classic PI. Like the PUA, however, all of them fall prey to the grand PI, which will become clear in the next subsections.

### 2.4.3 The Relative PI

Like Stanford (2006), Mario Alai (2017) develops a new PI to get around the first problem with the classic PI. His new PI states that "even in the past empirical knowledge and scientific methodology had improved steadily: for instance, they had improved a lot from 100 A.D. to 1700 A.D., yet many wrong theories were still held at that date, and even thereafter" (Alai, 2017, p. 3282). To put it differently, obsolete theories, although better than their forerunners, were thrown out, so their current successors, although better than their forerunners, will also be thrown out. Both the premise and the conclusion are relative statements. For this reason, I call this PI "the relative PI" (Park, 2019c, pp. 95–96). Its counterpart that I call "the absolute PI" coincides with the classic PI. It asserts that obsolete theories, even if successful, were overturned, so their current successors, even if successful, will also be overturned. Note that the premise and the conclusion each have an absolute form.

Is the relative PI cogent? It is not cogent because there are counterexamples to its inferential schema (Park, 2019c, pp. 98–100). Consider the inference that water did not boil when the temperature rose from 90 °C to 95 °C, so water will not boil when the temperature rises from 95 °C to 100 °C. The premise of this inference is true, while the conclusion is false. Such counterexamples open up the possibility that although the premise of the relative PI is true, its conclusion might be false. The counterexamples also point to the existence of what I call "a flipping point" (Park, 2017c, pp. 217–218). A flipping point is the point at which unobserved instances become dissimilar to observed instances. For example, 100 °C is a flipping point for water. Water does not boil at 95 °C. As its temperature rises, however, it reaches the point at which it starts to boil. Similarly, a scientific theory might have a flipping point. It is false at a certain degree of success, but as its degree of success rises, it might reach the point at which it becomes true.

On second thought, however, it is otiose for me to expose this intrinsic problem with the relative PI, for there is a more convenient means of refuting it. It is a successor to the classic PI intended to overcome the first intrinsic problem with the classic PI, just as the PUA is a successor to the classic PI intended to overcome the shortcoming of the classic PI. Consequently, the relative PI is vulnerable to the grand PI, just as the PUA is vulnerable to the grand PI. The grand PI says that since the classic PI is infinitely problematic, its successors, including the relative PI, are also infinitely problematic.

Any new PI intended to get around an intrinsic problem with the classic PI falls prey to the grand PI, according to which since the classic PI is infinitely problematic, its successors, including the new PI, are also infinitely problematic. The new PI is fated to be infinitely problematic, so we are entitled to reject it even before it is born. So what? Even before Alai (2017) developed the relative PI, we were entitled to reject it, although we did not know that we were. Let me now turn to another successor to the classic PI that suffers a similar fate.

### *2.4.4 The PI over Realists*

Like Stanford (2006) and Alai (2017), Wray (2013) and Nickles (2017) have developed a new PI to overcome the realist objection that newer theories are better than older ones. Wray states that just as what looked new to our ancestors do not look new to us, so "what looks new to us will not look so new to our offspring" (Wray, 2013, p. 4327). Analogously, Nickles asks, "Why should we think that today's best science is true when past scientists believed the same of their science—which we reject today as badly wrong?" (Nickles, 2017, p. 153). These philosophers' PI can be analyzed with the use of $T_1$, $T_2$, and $T_3$, where $T_1$ and $T_2$ are obsolete theories and $T_3$ is a current theory. Realists of $T_2$ believed that $T_2$ was true on the grounds that it was empirically more successful than $T_1$, but their beliefs about $T_2$ turned out to be false. Similarly, realists of $T_3$ believe that $T_3$ is true on the grounds that it is empirically more successful than $T_2$, but their beliefs about $T_3$ will also turn out to be false. I call this PI "the PI over realists" (Park, 2017b, p. 618).

How does the PI over realists relate to the relative PI discussed in Sect. 2.4.3 above? As far as I can tell, the PI over realists entails the relative PI. Consider again $T_1$, $T_2$, and $T_3$. The premise of the PI over realists states that realists of $T_2$ believed that $T_2$ was true on the grounds that it was empirically more successful than $T_1$, but their beliefs about $T_2$ turned out to be false. This premise entails the premise of the relative PI that $T_2$, although better than $T_1$, was thrown out. The conclusion of the PI over realists states that realists of $T_3$ believe that $T_3$ is true on the grounds that it is empirically more successful than $T_2$, but their beliefs about $T_3$ will also turn out to be false. This conclusion entails the conclusion of the relative PI that $T_3$, although better than $T_2$, will also be thrown out.

How does the PI over realists relate to the PUA? There are similarities and differences between them. The similarities are that they are both successors to the classic PI and that they are both constructed to overcome the first problem with the classic PI. The differences between them are that the PI over realists is directed at realists, whereas the PUA is directed at scientists, and that the PI over realists does not appeal to the notion that there are infinitely many alternatives, whereas the PUA does so.

What are we to make of the PI over realists? It entails the relative PI, and thus it inherits the intrinsic problem with the relative PI, viz., it overlooks the possible existence of a flipping point. However, it is otiose to raise such an objection to the PI over realists. The PI over realists is a successor to the classic PI, just as the PUA is a successor to the classic PI. Consequently, the PI over realists falls victim to the grand PI, according to which since the classic PI is infinitely problematic, its successors, including the PI over realists, are also infinitely problematic.

### 2.4.5 *The Epistemic PI*

Park (2019d, p. 284) distinguishes between semantic and epistemic PIs. The semantic PI states that current theories are false because their precursors were false, and it is identical with the classic PI. By contrast, the epistemic PI states that current theories are unwarranted because their precursors were unwarranted. Pessimists might construct the epistemic PI in response to my criticism in Sect. 2.2 of this chapter that if they retreat to the Humean argument, they cannot explain the success of science and they cannot accept instrumentalism. The epistemic PI enshrines Hume's uniformity principle, and thus it is not self-defeating for supporters of the epistemic PI to give antirealist explanations and accept instrumentalism.

How do I assess the epistemic PI? It inherits the intrinsic problems with the classic PI. For example, the first intrinsic problem with the classic PI applies just as well to the epistemic PI. Current theories are better than past ones. Consequently, it is a fallacious inference that since the latter were unjustified, the former are also unjustified. In addition, it falls prey to the grand PI, according to which since the classic PI has infinitely many problems, its successors, including the epistemic PI, also have infinitely many problems.

This criticism against the epistemic PI presupposes that the epistemic PI is a successor to the classic PI. But what if the epistemic PI is not a successor to the classic PI? What if the epistemic PI is instead the predecessor to the classic PI? Imagine that unbeknownst to us, someone had constructed the epistemic PI in the nineteenth century. Under this condition, the epistemic PI would not fall prey to the grand PI. All PIs, including the grand PI, involve the temporal directionality, i.e., they make inferences from older to newer ideas, but not from newer to older ideas. Therefore, the refutation of the epistemic PI would require that we should expose the intrinsic problems with it.

This conclusion accords with what I call "the first-mover advantage" (Park, 2018, p. 13). The first idea in a domain, whatever it is, enjoys the first-mover advantage, i.e., it cannot fall victim to a PI. For example, suppose that some scientists are first-movers, i.e., that they propose a theory for the first time in a new scientific field. Under these circumstances, it would be wrong to say that since its predecessor was abandoned, it will also be abandoned. The only way to refute it is to expose its intrinsic problems. To use an analogy, the only way to tip over the first domino in a succession of dominos is to push it.

By contrast, suppose that scientists propose a theory in an old field of science in which another theory has been accepted and rejected, i.e., their theory has a false predecessor. Under these circumstances, their theory would fall prey to a PI, and thus we do not need to expose any intrinsic problem with it to reject it. To take up the analogy again, if the first domino fell over, we do not need to push the second domino. The second domino will fall over whether we push it or not.

It is much more difficult to refute a scientific theory by exposing its intrinsic problems than to refute it by running a PI against it (Park, 2018, p. 13). For example, it took about 1200 years for scientists to expose the intrinsic problems with the Ptolemaic theory (Fahrbach, 2011a, pp. 141–142). However, it takes less than a minute to run the classic PI against the Copernican theory. The same goes for the PIs themselves. It is much harder to refute a first-order PI by exposing its intrinsic problems than to refute it by running a second-order PI against it. It took several decades for realists to refute the classic PI by exposing its four intrinsic problems. However, it takes less than a minute to run the grand PI against the PUA.

The grand PI is the first second-order PI in the history of PIs, i.e., it does not have a predecessor. It follows that it enjoys the first-mover advantage, i.e., the only way to refute it is to expose its intrinsic problems. If you think that the first-mover advantage is illusory, you are not a pessimist, and you cannot run any PI against scientific theories, scientists, or realists. Moreover, you are encouraged to add the notion of the first-mover advantage to the list of the intrinsic problems with the classic PI that are offered in Sect. 2.4.1 of this chapter.

### *2.4.6 The Problem of Misleading Evidence*

Greg Frost-Arnold (2019) has constructed the problem of misleading evidence (PME), which he intends to be a successor to the PUA. The PME concerns scientists' evidence for scientific theories. Its premise is that "Past scientists often faced misleading total evidence sets," and its conclusion is that "Current scientists often face misleading total evidence sets" (Frost-Arnold, 2019, p. 912). For example, Newton's evidence for his theory was misleading, and thus Albert Einstein's evidence for his theory is also misleading.

Frost-Arnold argues that there are some historical episodes that the PME can handle, but that the PUA cannot. For example, Ptolemy was exposed to the heliocentric theory, but he rejected it and accepted the geocentric theory (Frost-Arnold, 2019, p. 911). It is right for proponents of the PME to say that Ptolemy accepted the geocentric theory over the heliocentric theory because his evidence was misleading. By contrast, it is wrong for advocates of the PUA to say that Ptolemy accepted the geocentric theory over the heliocentric theory because he was unable to conceive of alternatives to the geocentric theory.

How do I appraise the PME? Overall, it is an original and impressive argument. Just as the PUA makes improvements on the classic PI, so the PME makes improvements on the PUA. These improvements mirror the improvements that scientific theories make over their forerunners. Consequently, realists should take the PME seriously.

The PME nevertheless falls victim to the grand PI, according to which since the classic PI and the PUA are infinitely problematic, its successors, including the PME, are also infinitely problematic. The PME has two precursors: the classic PI and the PUA. By contrast, the PUA has only one precursor: the classic PI. Consequently, the PME is more likely to be infinitely problematic than the PUA. Grand pessimists do not specify any intrinsic problem with the PME to argue that the PME is infinitely problematic any more than proponents of the PME specify any intrinsic problem with the evidence for Einstein's theory to argue that its evidence is misleading. Grand pessimists only point out that the PUA is infinitely problematic, just as proponents of the PME only point out that the evidence for Newton's theory was misleading. Therefore, it would be a double standard for proponents of the PME to request that grand pessimists specify an intrinsic problem with the PME.

Proponents of the PME might counter that the PME is more successful than the PUA, given that some historical episodes can be accommodated by the PME, but not by the PUA. Therefore, it is a fallacious inference that since the PUA is infinitely problematic, the PME is also infinitely problematic. However, it is obvious how grand pessimists would defuse this argument. The PUA was better than the classic PI, but it turned out to be infinitely problematic. It follows that the PME will also turn out to be infinitely problematic, although it is better than the PUA.

Grand pessimists are happy to see the PME in the literature. Before the PME was born, grand pessimists predicted that the PUA would "be replaced by a future pessimistic induction hitherto unconceived" (Park, 2018, p. 12). Given that the PME was just introduced into the literature, it will be around for a while. As predicted, however, it "has infinitely many problems, just as its two forerunners do" (Park, 2018, p. 12). Consequently, it will be displaced by a successor. We do not yet know the name of this successor, but we will, just as we did not know the name "the PME," but we do now.

The following dilemma can be constructed vis-à-vis the PME. Scientific theories have been replaced by their successors. A similar pattern can be found in the classic PI, the PUA, and the PME. On the one hand, if the historical pattern of scientific theories is a good indication that the evidence for a current theory is misleading, the historical pattern for the three first-order PIs is also a good indication that the PME is infinitely problematic. On the other hand, if the historical pattern for scientific theories is not a good indication that the evidence for the current theory is misleading, then the PME is fallacious because it asserts that the historical pattern of scientific theories is a good indication that the evidence for a current theory is misleading. On either horn of the dilemma, we should reject the PME.

The dilemma has the effect of strengthening the grand PI, given that the grand PI predicted, before this dilemma was even constructed, that the classic PI, the PUA, and their successors have infinitely many hidden intrinsic problems. The grand PI specifies only four of the infinitely many intrinsic problems with the classic PI and none of the infinitely many intrinsic problems with the PUA and the PME. The dilemma just revealed one of the hitherto hidden intrinsic problems with the PME, thereby supporting the prediction of the grand PI that the successors to the classic PI have infinitely many hidden intrinsic problems.

The dilemma bolsters the grand PI, and it does not take the place of the grand PI. The grand PI offers a much stronger challenge to the PME than the dilemma. The dilemma only implies that there is an intrinsic problem with the PME. By contrast, the grand PI claims that there are infinitely many intrinsic problems with the PME. Moreover, the grand PI asserts that there are infinitely many first-order PIs, that the PME will be displaced by a hitherto unconceived PI, and that they are all infinitely problematic. Such assertions are not found in the dilemma.

Pessimists might object that the grand PI commits two hasty generalizations (Park, 2018, p. 14). The first hasty generalization can be found in the inference from the existence of the four intrinsic problems with the classic PI to the existence of infinitely many intrinsic problems with it. There is an infinitely wide gap between the premise and the conclusion. The second hasty generalization can be found in the inference from the existence of the classic PI and the PUA to the existence of infinitely many hitherto unconceived first-order PIs. There is also an infinitely wide gap between the premise and the conclusion.

However, this objection to the grand PI comes at a cost for advocates of the PUA. There is also an infinitely wide gap between the conclusion and the premise of the PUA. The premise is that past and present scientists could not ideate alternatives to the theories of their times, and the conclusion is that current theories have infinitely many alternatives. Stanford offers the following list of successive theories to support the premise:

**Stanford's List.**

from elemental to early corpuscularian chemistry to Stahl's phlogiston theory to Lavoisier's oxygen chemistry to Daltonian atomic and contemporary chemistry,

from various versions of preformationism to epigenetic theories of embryology,

from the caloric theory of heat to later and ultimately contemporary thermodynamic theories,

from effluvial theories of electricity and magnetism to theories of the electromagnetic ether and contemporary electromagnetism,

from humoral imbalance to miasmatic to contagion and ultimately germ theories of disease,

from eighteenth century corpuscular theories of light to nineteenth century wave theories to the contemporary quantum mechanical conception,

from Darwin's pangenesis theory of inheritance to Weismann's germ-plasm theory to Mendelian and then contemporary molecular genetics,

from Cuvier's theory of functionally integrated and necessarily static biological species and from Lamarck's autogenesis to Darwin's evolutionary theory (Stanford, 2006, pp. 19–20).

Note that Stanford takes the existence of a few theories of heat to be evidence for the existence of the indefinitely or infinitely many theories of heat. Therefore, we can conclude that the PUA and the grand PI rise and fall together with respect to the issue of infinity (Park, 2018, p. 14).

In light of this reply, advocates of the PUA might retreat to the modest thesis that the number of unconceived competitors is finite, but they might insist that even if the number is finite, the PUA poses a serious threat to realism, provided that the number is large enough. For example, if the kinetic theory has one hundred alternatives, it is fated to be overthrown.

In response, however, grand pessimists can make a similar move. They can argue that even if the number of problems with the classic PI, the PUA, and the PME is finite, the grand PI poses a serious threat to the first-order PIs, provided that the number of problems with them is large enough. For example, if the PME has one hundred problems, it is extremely problematic and fated to be ousted. Grand pessimists can also argue that even if the number of first-order PIs is finite, the grand PI poses a serious threat to the PME, provided that the number is large enough. For example, if the PME has one hundred unconceived alternatives, it is fated to be overturned. In sum, if infinity is a dispensable part of the PUA, it is also a dispensable part of the grand PI.

Now, proponents of the PME might be tempted to raise the following four objections to the grand PI. (i) Since the classic PI, the PUA, and the PME are infinitely problematic, the grand PI is also infinitely problematic. (ii) The status of the PME is independent of the status of the classic PI and the PUA. (iii) The PME is not a successor to the PUA, and thus the pessimistic inference from the PUA to the PME is illegitimate. (iv) Philosophy and science are different intellectual enterprises, and thus scientific ideas are susceptible to a PI, while philosophical ideas, including the PIs, are not. Objections similar to these have already been preempted (Park, 2018, pp. 13–17). For the sake of space, I explore only the following possible dialectic between proponents of the PME and grand pessimists.

Proponents of the PME might object that since the classic PI, the PUA, and the PME are infinitely problematic, the grand PI is also infinitely problematic. In response, grand pessimists would raise a similar objection to the PME, i.e., because the evidence for the caloric theory was misleading, the evidence for the kinetic theory is misleading, and the evidence for the PME is also misleading. Adherents of the PME would protest that the PME is not a successor to the kinetic theory, so it is wrong to infer the downfall of the PME from that the caloric and kinetic theories. However, grand pessimists would make a similar protest that it is wrong to infer the downfall of the grand PI from that of the classic PI, the PUA, and the PME.

To return to our analogy, the caloric and kinetic theories are like a succession of dominos. If the caloric theory falls down, the kinetic theory will also fall down. By contrast, the PME is not another domino behind the kinetic theory. It is rather a domino above the two dominos. Consequently, it would not fall down just because the two dominos below it fall down. Grand pessimists would say the same about the grand PI. The classic PI, the PUA, and the PME are a succession of dominos. If the classic PI falls down, the PUA and the PME fall down. By contrast, the grand PI is not another domino behind the PME. It is rather a domino above the three dominos. Therefore, it would not fall down just because the three dominos below it fall down.

Let me add that the PME feeds off the downfall of scientific theories. The more scientific theories fall down due to their misleading evidence, the stronger the PME becomes. Analogously, the grand PI feeds off the downfall of first-order PIs. The more first-order PIs fall down, the stronger the grand PI becomes. Therefore, we should infer not the downfall but the fortification of the grand PI from the downfall of the classic PI, the PUA, and the PME.

Let me now turn to a possible move that advocates of the PUA and the PME might make. They might argue that the PUA and the PME should not be thought to replace the classic PI in the same way that the Copernican theory replaced the Ptolemaic one. Rather, the PUA and the PME should be understood as embedding the classic PI, i.e., the classic PI should be understood as reducible to the PUA and the PME in the same way that thermodynamics is reducible to statistical mechanics. Thus, the relation between the PUA and the PME, on the one hand, and the classic PI, on the other hand, is more like the relation between thermodynamics and statistical mechanics than the relation between the Ptolemaic and Copernican theories.

Before evaluating this intriguing move, we should first be clear about what it means to say that thermodynamics is reducible to statistical mechanics. It means, among other things, that thermodynamics is deducible from statistical mechanics (Nagel, 1979, p. 345). In order for this deduction to be possible, an important condition has to be met, viz., the two theories must be descriptions of the same objects. Stated differently, their key terms must refer to the same things. If they referred to different things, thermodynamics would not be deducible from statistical mechanics. For example, "heat" should refer to the kinetic energy of molecules. If it does not, thermodynamics cannot be reduced to statistical mechanics.

Is the classic PI reducible to the PUA? Answering this question requires that we investigate whether they talk about the same thing, just as thermodynamics and statistical mechanics do. In my view, they describe different objects. As noted in Sect. 2.4 of this chapter, the classic PI concerns scientific theories, while the PUA concerns scientists. It is problematic to suggest that theories turn out to be theorists, so theories are numerically identical with theorists, in the same way that heat turned out to be the kinetic energy of molecules, so heat is numerically identical with the kinetic energy of molecules. Since the classic PI and the PUA concern distinct objects, the classic PI is irreducible to the PUA.

Is the classic PI reducible to the PME? In my view, they make assertions about distinct objects. The classic PI concerns scientific theories, whereas the PME concerns the evidence for scientific theories. It is problematic to contend that scientific theories turn out to be the evidence for them, so scientific theories and the evidence for them are the same things, just as heat turned out to be the kinetic energy of molecules, so they are the same thing. Since the classic PI and the PME concern distinct objects, the classic PI is irreducible to the PME.

Moreover, the PUA is irreducible to the PME. The PUA concerns scientists, whereas the PME concerns the evidence for theories. It is problematic to claim that scientists turn out to be the evidence for scientific theories, so scientists are numerically identical with the evidence. Since the PUA and the PME concern distinct targets, the PUA is not reducible to the PME. I suppose that advocates of the PUA would find this conclusion agreeable.

In addition, why not say the same thing about scientific theories? That is, why not say that just as the classic PI is reducible to the PUA and the PME, so current scientific theories will be reducible to future scientific theories? It is a double standard to say that the classic PI is reducible to the PUA and the PME, but that current scientific theories will not be reducible to future scientific theories.

Frost-Arnold would argue that the PME is not a successor to the PUA. He claims that although there are historical cases that the PME can explain but the PUA cannot, there are also historical cases that "the PUA explains at least as well as the PME" (Frost-Arnold, 2019, p. 915). Since the PME is not a successor to the PUA, he might continue, the PME does not fall prey to the grand PI.

My response to this argument is to say the same thing about scientific theories. Although there are phenomena that present theories can explain but past theories cannot, there are also phenomena that past theories can explain at least as well as present theories. Since present theories are not successors to past theories, they do not fall victim to the classic PI. The same holds for present scientists and the evidence for present theories, i.e., present scientists are not successors to past scientists, and the evidence for present theories is not the successor to the evidence for past theories. It follows that the classic PI, the PUA, and the PME are all fallacious, which coincides with what the grand PI asserts.

### 2.4.7 Proportional Pessimism

In this chapter, we discussed six first-order PIs: the classic PI, the PUA, the relative PI, the PI over realists, the epistemic PI, and the PME. The classic PI is the common ancestor of the five first-order PIs. Of the five first-order PIs, which is most likely to be infinitely problematic? The PME has two predecessors, viz., the classic PI and the PUA. By contrast, the PUA, the relative PI, the PI over realists, and the epistemic PI each have only one predecessor, viz., the classic PI. Therefore, the PME is most likely to be infinitely problematic, and the PUA, the relative PI, the PI over realists, and the epistemic PI are equally likely to be infinitely problematic.

An interesting corollary of the foregoing discussion is that if pessimists aim to develop a successor to an existing first-order PI, they are encouraged to devise a successor to the classic PI rather than a successor to the PME. After all, a successor to the classic PI would have only one predecessor, viz., the classic PI. By contrast, a successor to the PME would have three predecessors: the classic PI, the PUA, and the PME. Therefore, the successor to the classic PI would have a better chance to succeed than the successor to the PME.

This conclusion goes hand in hand with what I call "proportional pessimism" (Park, 2016, p. 836). Proportional pessimism is the idea that the probability that a theory will be abandoned is proportional to the number of its predecessors. That is, the more false precursors a theory has, the more probable it is that it will be abandoned. Suppose that we have two scientific theories in different domains. One has one false predecessor, and the other has two false predecessors. In such a scenario, the classic PI implies that the latter is more probably false than the former. It is for this reason that first-order pessimists strive for a long list of obsolete theories. For example, Stanford's (2006, pp. 19–20) list cited in Sect. 2.4.6 of this chapter is longer than Laudan's (1981, p. 33) list cited in Sect. 1.3 of Chap. 1. According to proportional pessimism, Stanford's list is more forceful than Laudan's list.

Readers might think that proportional pessimism is an outrageous thesis. Suppose that physicists are not happy with Einstein's mechanics and that they aim to develop a new mechanics. If they have such an aim, they should develop a new mechanics that improves upon Einstein's mechanics, not upon Aristotle's mechanics. Newton's mechanics improves upon Aristotle's mechanics, and Einstein's mechanics improves upon Newton's mechanics. Therefore, a successor to Einstein's mechanics would be better than a successor to Aristotle's mechanics. Similarly, suppose that first-order pessimists are not happy with the PME and that they aim to develop a new first-order PI. If they have such an aim, they should develop a new PI that improves upon the PME, but not upon the classic PI. The PUA improves upon the classic PI, and the PME improves upon the PUA. Consequently, a successor to the PME would be better than a successor to the classic PI. If readers think so, however, they are encouraged to add proportional pessimism to the list of the intrinsic problems with the classic PI in Sect. 2.4.1 of this chapter.

## 2.5 Summary

This chapter discusses one second-order PI and six first-order PIs. The second-order PI is the grand PI, and the six first-order PIs are the classic PI, the PUA, the relative PI, the PI over realists, the epistemic PI, and the PME. The grand PI asserts that since the classic PI is infinitely problematic, its successors, which are infinitely many, are also infinitely problematic. The grand PI spells doom for the six first-order PIs and their successors. The four intrinsic problems with the classic PI are the source of trouble for the six first-order PIs and their successors.

Thanks to the grand PI, realists do not need to worry about a new first-order PI that pessimists will put forward in the future. Realists are entitled to predict, even before it is born, that it is infinitely problematic. After all, the fact that it has a few predecessors that are infinitely problematic makes it automatically and infinitely problematic as a matter of induction. It might initially appear to be powerful, but it will, in fact, be powerless, and thus it will not negatively affect realism.

I challenge pessimists to refute the grand PI without thereby also refuting the six first-order PIs. I predict that their criticisms against the grand PI will inevitably backfire on the six first-order PIs, thereby only lending greater credence to the grand PI. I dare say that once their first attempt to refute the grand PI fails, their second attempt will also fail as a matter of induction and that they will be trapped in an infinite chain of failures to refute the grand PI. As a result, the grand PI will only be fortified to an infinite degree. For this reason, their first attempt to refute the grand PI had better succeed. This chapter shows, I hope, that running the grand PI is the most convenient way to defend realism against the six first-order PIs and their successors.

## References


Alai, M. (2017). Resisting the historical objections to realism: Is Doppelt's a viable solution? *Synthese, 194*(9), 3267–3290.

Asay, J. (2019). Going local: A defense of methodological localism about scientific realism. *Synthese, 196*(2), 587–609.

Bird, A. (2007). What is scientific progress? *Noûs, 41*(1), 92–117.

Dellsén, F. (2019). Should scientific realists embrace theoretical conservatism? *Studies in History and Philosophy of Science, 76*, 30–38.

Devitt, M. (2011). Are unconceived alternatives a problem for scientific realism? *Journal for General Philosophy of Science, 42*(2), 285–293.

Doppelt, G. (2007). Reconstructing scientific realism to rebut the pessimistic meta-induction. *Philosophy of Science, 74*(1), 96–118.

Doppelt, G. (2014). Best theory scientific realism. *European Journal for Philosophy of Science, 4*(2), 271–291.

Fahrbach, L. (2011a). How the growth of science ends theory change. *Synthese, 180*(2), 139–155.

Fahrbach, L. (2011b). Theory change and degrees of success. *Philosophy of Science, 78*(5), 1283–1292.

Frost-Arnold, G. (2019). How to be a historically motivated antirealist: The problem of misleading evidence. *Philosophy of Science, 86*(5), 906–917.

Hume, D. (1888/1978). In L. A. Selby-Bigge & P. H. Nidditch (Eds.), *A treatise of human nature*. Oxford University Press.

Kitcher, P. (1993). *The advancement of science: Science without legend, objectivity without illusions*. Oxford University Press.

Kuhn, T. (1962/1970). *The structure of scientific revolutions*. University of Chicago Press.

Ladyman, J. (1999). Review. A novel defense of scientific realism. J. Leplin. *The British Journal for the Philosophy of Science, 50*(1), 181–188.

Lange, M. (2002). Baseball, pessimistic inductions and the turnover fallacy. *Analysis, 62*(4), 2881–2885.

Laudan, L. (1977). *Progress and its problems: Towards a theory of scientific growth*. University of California Press.

Laudan, L. (1981). A confutation of convergent realism. *Philosophy of Science, 48*(1), 19–49.

Leplin, J. (1997). *A novel defense of scientific realism*. Oxford University Press.

Lyons, T. (2017). Epistemic selectivity, historical threats, and the non-epistemic tenets of scientific realism. *Synthese, 194*(9), 3203–3219.

Lyons, T. (2018). Four challenges to epistemic scientific realism. *Spontaneous Generations: A Journal for the History and Philosophy of Science, 9*(1), 146–150.

Mizrahi, M. (2013). The pessimistic induction: A bad argument gone too far. *Synthese, 190*(15), 3209–3226.

Mizrahi, M. (2015). Historical inductions: New cherries, same old cherry-picking. *International Studies in the Philosophy of Science, 29*(2), 129–148.

Mizrahi, M. (2016). The history of science as a graveyard of theories: A philosophers' myth. *International Studies in Philosophy of Science, 30*(3), 263–278.

Mizrahi, M. (2020). *The relativity of theory: Key positions and arguments in the contemporary scientific realism/antirealism debate*. Springer.

Musgrave, A. (1985). Realism vs. constructive empiricism. In P. Churchland & C. Hooker (Eds.), *Images of science: Essays on realism and empiricism* (pp. 197–221). University of Chicago Press.
Nagel, E. (1979). *The structure of science: Problems in the logic of scientific explanation*. Hackett Publishing Company.
Nickles, T. (2017). Cognitive illusions and nonrealism: Objections and replies. In E. Agazzi (Ed.), *Varieties of scientific realism: Objectivity and truth in science* (pp. 151–163). Springer.
Park, S. (2011a). Coherence of our best scientific theories. *Foundations of Science, 16*(1), 21–30.
Park, S. (2011b). A confutation of the pessimistic induction. *Journal for General Philosophy of Science, 42*(1), 75–84.
Park, S. (2016). Refutations of the two pessimistic inductions. *Philosophia, 44*(3), 835–844.
Park, S. (2017a). Problems with using evolutionary theory in philosophy. *Axiomathes, 27*(3), 321–332.
Park, S. (2017b). Why should we be pessimistic about antirealists and pessimists? *Foundations of Science, 22*(3), 613–625.
Park, S. (2017c). The uniformity principle vs. the disuniformity principle. *Acta Analytica, 32*(2), 213–222.
Park, S. (2018). The grand pessimistic induction. *Review of Contemporary Philosophy, 17*, 7–19.
Park, S. (2019a). In defense of realism and selectivism from Lyons's objections. *Foundations of Science, 24*(4), 605–615.
Park, S. (2019b). New objections to the problem of unconceived alternatives. *Filosofia Unisinos, 20*(2), 138–145.
Park, S. (2019c). The absolute and relative pessimistic inductions. *PRO, 95*, 94–104.
Park, S. (2019d). The descriptive and normative versions of scientific realism and pessimism. *Filozofia: Journal for Philosophy, 74*(4), 278–290.
Poincaré, H. (1905/1952). *Science and hypothesis*. Dover.
Psillos, S. (1997). How not to defend constructive empiricism: A rejoinder. *The Philosophical Quarterly, 47*(188), 369–372.
Psillos, S. (1999). *Scientific realism: How science tracks truth*. Routledge.
Putnam, H. (1978). *Meaning and the moral sciences*. Routledge & K. Paul.
Stanford, K. (2006). *Exceeding our grasp: Science, history, and the problem of unconceived alternatives*. Oxford University Press.
van Fraassen, B. (1989). *Laws and symmetry*. Oxford University Press.
Vickers, P. (2017). Understanding the selective realist defence against the PMI. *Synthese, 194*(9), 3221–3232.
Worrall, J. (1989). Structural realism: The best of both worlds. *Dialectica, 43*(1–2), 99–124.
Wray, B. (2008). The argument from underconsideration as grounds for anti-realism: A defence. *International Studies in the Philosophy of Science, 22*(3), 317–326.
Wray, B. (2010). Selection and predictive success. *Erkenntnis, 72*(3), 365–377.
Wray, B. (2011). *Kuhn's evolutionary social epistemology*. Cambridge University Press.
Wray, B. (2012). Epistemic privilege and the success of science. *Noûs, 46*(3), 375–385.
Wray, B. (2013). The pessimistic induction and the exponential growth of science reassessed. *Synthese, 190*(18), 4321–4330.
Wray, B. (2015). Pessimistic inductions: Four varieties. *International Studies in the Philosophy of Science, 29*(1), 61–73.

# Chapter 3
# Critiques of Scientific Antirealism

**Abstract** Antirealists are subject to the pessimistic induction over antirealists. Antirealists cannot explain events in terms of T due to Moore's paradox, the problem of disconcerting questions, and the problem of deceptive speech acts. Antirealists will not and cannot predict theoretical events. Antirealists' beliefs in their own positive theories are unjustifiable. Moreover, epistemic reciprocalists would not believe antirealists' theories. Antirealists might reply that they believe that T is empirically adequate or that there is a gap between what they say and what they believe. These replies are all problematic.

**Keywords** Constructive empiricism · Deceptive speech · Disconcerting questions · Epistemic disadvantages · Epistemic reciprocalism · Gap antirealism · Moore's paradox · Pessimistic induction · Pragmatic disadvantages

## 3.1 Introduction

Antirealism is any position that opposes realism. Antirealists typically make the following claims: (i) Our best theories are false. (ii) They will be abandoned. (iii) They are unwarranted. (iv) We cannot justifiably believe them. (v) It is unreasonable to believe them. (vi) It is rational to disbelieve them. (vii) We should be skeptical that theoretical entities exist. (viii) Our best theories are merely useful. (ix) Our best theories are merely empirically adequate (EA). Anyone who makes such claims is an antirealist.

The participants in the realism debate have been largely concerned with whether realism is tenable. The basic format of the debate between realists and antirealists has been that antirealists criticize realism and realists defend it from antirealists' criticisms. In my view, the disputants have largely neglected the question of whether antirealism is tenable, when it deserves an equal amount of attention. It deserves such attention because it is an alternative to realism. Accordingly, I make explicit what its advantages and disadvantages are in this chapter. It is clear what its advantage is: it runs less epistemic risk than realism. As far as I know, antirealism has no


S. Park, *Embracing Scientific Realism*, Synthese Library 445, https://doi.org/10.1007/978-3-030-87813-9_3

other advantage, while it has several epistemic and pragmatic disadvantages, which will become clear in this chapter.

The outline of this chapter is as follows. In Sects. 3.2, 3.3, and 3.4, I unfold the pessimistic induction (PI) over antirealist explanations of the success of science, the PI over pessimists, and the PI over antirealists, respectively. In Sect. 3.5, I disclose three problems that arise when antirealists invoke T, a scientific theory, for an explanatory purpose. The three problems are Moore's paradox, the problem of disconcerting questions, and the problem of deceptive speech acts. In Sect. 3.6, I argue that antirealists will not and cannot predict unobservable events. In Sect. 3.7, I expose two epistemic disadvantages with antirealism. In Sect. 3.8, I flesh out a position that I call "epistemic reciprocalism" in order to expose other epistemic disadvantages and some pragmatic disadvantages with antirealism. In Sect. 3.9, I criticize antirealists' response that they believe that T is EA. In Sect. 3.10, I explore whether my criticisms against antirealism apply to Bas van Fraassen's (1980) constructive empiricism. In Sect. 3.11, I criticize antirealists' possible reply that what they believe differs from what they say, so my criticisms against what they say do not refute what they believe.

## 3.2 The PI Over Antirealist Explanations

The no-miracles argument (NMA) asserts that the realist explanation is the best of all the conceived ones. Some philosophers take issue with this assertion, advancing antirealist alternatives and challenging realists to show that the realist explanation is superior to those antirealist alternatives. There are nine such alternatives (van Fraassen, 1980; Laudan, 1984; Fine, 1986, 1991; Musgrave, 1988; Brown, 1994; Kukla, 1998; Stanford, 2000; Lyons, 2003). Seungbae Park (2014a) exposes the intrinsic problems with each of the eight antirealist alternatives, but not with the latest one, which is proposed by Lyons (2003), and then predicts that it will suffer a similar fate. In short, I have constructed the PI over antirealist explanations.

What do the nine antirealist alternatives say? What are their problems? I refer readers to Park (2014a) for answers to these questions. Let me say here, though, that I have already discussed some of them in Chap. 1. In Sect. 1.9 of Chap. 1, I argued that it is circular to explain success in terms of empirical adequacy. In Sect. 1.10 of Chap.r 1, I argued that surrealism fails to recognize scientists' originality. In Sect. 1.11 of Chap. 1, I argued that predictive similarity is not an adequate explanans for success.

Antirealists might object that science is vulnerable to a PI, but philosophy is not, and thus the PI over antirealist explanations is fallacious. However, Park (2014a, pp. 17–19, 2018a, pp. 16–17) has already made several critical comments on this possible objection from antirealists. Let me mention only two of them here. First, the history of science tells us that observational data were relatively stable, although theories have been relatively unstable. Scientific theories are directly connected with observational data, whereas philosophical theories are indirectly connected with them. Therefore, philosophical theories are more vulnerable to a PI than scientific theories. Second, what about thinkers, such as Richard Healey and Sean Carroll, who do research on the borderline between science and philosophy? It sounds implausible to say that their theories are *half* vulnerable to a PI.

How could antirealists get around the PI over antirealist explanations? The best way, I believe, is to argue that my (Park, 2014a) criticisms against the eight antirealist alternatives are flawed. For example, they might show that contrary to my opinion that predictive similarity is not an adequate explanans, predictive similarity is, in fact, an adequate explanans. Since the premise of the PI over antirealist explanations is false, we do not need to accept the pessimistic prediction that Lyons's (2003) alternative has intrinsic problems.

I welcome antirealists to engage with Park (2014a) and expose intrinsic flaws with my criticisms against the eight antirealist alternatives. However, I point out that their first attempt to rebut my criticisms should succeed. If their first attempt fails, an inductive rationale will emerge "for thinking that their second attempt will also fail" (Park, 2017a, p. 622). It is wrong to think that scientists' failure constitutes an inductive rationale for predicting that there will be more failures, but that philosophers' failure does not constitute a reason for predicting that there will be more failures.

This quagmire for antirealists reminds us of the epistemic imperative: "Act only on an epistemic maxim through which you can at the same time will that it should become a universal one" (Park, 2018b, p. 441). Consider the epistemic maxim, "Run a PI against others." If pessimists do not will this maxim to become a universal one, they should not run the six first-order PIs against scientists, realists, and scientific theories, which were unfolded in Chap. 2. If they will the maxim to become a universal one, they would have to agree with me that philosophers' failure constitutes a reason for predicting that there will be more failures.

## 3.3 The PI Over Pessimists

Ludwig Fahrbach (2011), Park (2011), and Moti Mizrahi (2013, 2016) distinguish between theories in the distant and recent past, as we noted in Sect. 2.4.1 of Chap. 2. They observe that the former were abandoned before the twentieth century, while the latter have not yet been abandoned, and that the number of the former is far lower than that of the latter.

This historical fact entitles us to construct the following PI over pessimists. Imagine that the pessimists of the early twentieth century expected that the theories of their time would be discarded. We can see from the viewpoint of the twenty-first century that they were wrong. The pessimists of the twenty-first century expect that the theories of their time will be disproven. However, our descendants of the twenty-second century will similarly see from the viewpoint of the twenty-second century that the pessimists of the twenty-first century are wrong (Park, 2017a, p. 623).

This PI over pessimists parallels Brad Wray's (2013) PI over realists. Wray asks us to imagine that the realists from the distant past believed the theories of their time. We can see from the viewpoint of the twenty-first century that their beliefs were false. Analogously, "what looks new to us will not look so new to our offspring" (Wray, 2013, p. 4327). To put it another way, our philosophical offspring will see from their viewpoint that contemporary realists' beliefs about current theories are false, as noted in Sect. 2.4.4 of Chap. 2.

The two preceding PIs—the PI over pessimists and the PI over realists—have opposing implications for the fate of current theories. The PI over pessimists implies that they are fated to be retained, while the PI over realists implies that they are fated to be discarded. The different implications stem from the different appeals to different groups of past theories. The PI over pessimists appeals to theories from the recent past, while the PI over realists appeals to those from the distant past. Both PIs operate under the same principle, viz., the uniformity principle.

Which PI is stronger, the PI over pessimists or the PI over realists? My answer is that the PI over pessimists is stronger than the PI over realists. We should not rely on the theories from the distant past, but rather should rely on those from the recent past, when assessing the fate of current theories. This is because current theories are more akin to the latter than to the former in terms of their capacity to explain and predict phenomena (Wray, 2015, p. 63; Park, 2019a, p. 97). To use an analogy, we should test a new drug on white mice rather than on snakes when predicting whether the drug will work on humans because humans are physiologically more akin to the former than to the latter.

The PI over pessimists radically differs from the six first-order PIs that we critically examined in Chap. 2: the classic PI, the PUA, the relative PI, the PI over realists, the epistemic PI, and the PME. The PI over pessimists was devised to support realism, while the six PIs have been devised to rebut realism. Moreover, the PI over pessimists is directed at proponents of the six PIs. Proponents of the six PIs are all pessimists who infer the falsehood of current theories from that of their predecessors. If they had advocated their PIs in the early twentieth century, their beliefs about scientists, theories, and realists would have been false, and hence their theoretical offspring's beliefs about scientists, theories, and realists would also be false.

## 3.4 The PI Over Antirealists

Antirealists have made many philosophical mistakes, as we have seen in Chaps. 1 and 2 and in Sect. 3.2 of this chapter. For example, van Fraassen (1989) advances the argument from a bad lot without being aware that it has the intrinsic flaw of drawing a normative conclusion from a possibility premise, as noted in Sect. 1.5 of Chap. 1. Kyle Stanford develops the PUA without being aware that it falls prey to the grand PI, as we have seen in Sect. 2.4.2 of Chap. 2. Antirealists' other mistakes are summarized in Park (2017b, pp. 31–32).

It is worth mentioning that James Ladyman et al. (1997) also make a philosophical mistake. They claim, "Certainly the realist takes an extra epistemic risk by believing the background theories to be (approximately) true rather than only empirically adequate" (Ladyman et al., 1997, p. 308). This claim is questionable in light of the classic PI (Park, 2014b, pp. 112–115). If the classic PI is correct, past and present theories are "empirically inadequate" (Park, 2001, p. 78; Lyons, 2003, p. 898). Obsolete theories ran into anomalies, which indicates that they were not EA. However, selective realists argue that obsolete theories were approximately true, given that their working assumptions were true (Psillos, 1999, p. 113). Admittedly, it is controversial whether they were approximately true or completely false. As Stanford observes, the disagreement between pessimists and selective realists "is simply a difference of style or taste in applying the expression 'approximately true' rather than a substantive disagreement between them" (Stanford, 2015, p. 875). By contrast, pessimists and selective realists can agree that past theories were not EA.

What can we conclude from antirealists' philosophical mistakes? We can construct the PI over antirealists, according to which "antirealists have made philosophical mistakes in the past, so they must be making philosophical mistakes now" (Park, 2017b, p. 31). The PI over antirealists is different from the PI over pessimists in Sect. 3.2 of this chapter in that the scope of the former is wider than that of the latter. "Antirealists" is a more inclusive term than "pessimists." All pessimists are antirealists, but not vice versa. For example, van Fraassen is an antirealist, but he is not a pessimist. Throughout his career, he has never advocated a PI, though he does not believe in current theories.[1] Thus, his philosophical mistakes reinforce the premise of the PI over antirealists, but not the premise of the PI over pessimists. Relatedly, he falls within the range of the PI over antirealists, but not within the range of the PI over pessimists.

In response, antirealists might conjure up the grand PI, i.e., they might argue that since the classic PI is infinitely problematic, its successors, including the PI over antirealists, are also infinitely problematic. In my view, however, this application of the grand PI is inadequate for the following two reasons. First, on closer examination, the demise of the classic PI supports rather than undermines the PI over antirealists. The demise of the classic PI means that pessimists made a theoretical mistake, and that their theoretical mistake supports the premise of the PI over antirealists, which is that antirealists have made theoretical mistakes. Since the demise of the classic PI supports the premise of the PI over antirealists, the conclusion of the PI over antirealists is more likely to be true. Keep in mind that the PI over antirealists feeds off antirealists' philosophical mistakes, including pessimists' philosophical mistakes. The more pessimists have made mistakes, the stronger the PI over antirealists is.

[1] Van Fraassen (1989, p. 171) chooses the English notion of rationality (which I discuss at greater length in Chapter 4) to establish that it is rational for an agent to believe and for another agent not to believe current theories, which implies that according to him, the classic PI does not deter realists from rationally believing current theories (Park, 2020a, pp. 3–5).

Second, the PI over antirealists is not a successor to the classic PI, and it is not designed to overcome a difficulty with the classic PI. Consequently, it is wrong to say that since the classic PI turned out to be incorrect, the PI over antirealists will also turn out to be incorrect. The PI over antirealists is fundamentally different from the PUA. The PUA is a successor to the classic PI, designed to get around a problem with the classic PI, just as the kinetic theory is a successor to the caloric theory, designed to get around a problem with the caloric theory. Therefore, it is legitimate to say that since the classic PI turned out to be incorrect, the PUA will also turn out to be incorrect. However, it is illegitimate to say that since the classic PI turned out to be incorrect, the PI over antirealists will turn out to be incorrect.

The PI over antirealists is the first idea in a separate domain that concerns antirealists. Accordingly, it does not have any predecessor, and it would be wrong to say that since its predecessor turned out to be incorrect, it would also turn out to be incorrect. It enjoys the first-mover advantage. The first idea in a domain is not subject to a PI, and the refutation of it requires that its intrinsic problems should be exposed, as we discussed in Sect. 2.4.5 of Chap. 2. Therefore, antirealists would have to expose the intrinsic problems of the PI over antirealists instead of running a second-order PI against it.

Antirealists might retort that all of the first ideas in all human intellectual enterprises thus far have been overturned, and hence the PI over antirealists, which is the first idea in the philosophical enterprise which concerns antirealists' ideas, will also be overturned (Park, 2014a, p. 19). My response to this PI over all first ideas is to ask whether the PI over all first ideas is a first idea or not. On the one hand, if it is a first idea, it will also be overturned, just as the other first ideas were overturned. On the other hand, if it is not, it will be overturned due to the demise of its predecessor(s), i.e., it will fall over like the other dominos in front of it. Therefore, it will be overturned whether it is a first idea or not.

## 3.5 Explanation

Many philosophers think that we can use T to explain events even if we do not believe it. Those philosophers are van Fraassen (1980, p. 12), Rasmus Winther (2009, p. 376), Gregory Dawes (2013, p. 68), Finnur Dellsén (2017, p. 249), and Healey (2019a, p. 43). For example, van Fraassen says that "acceptance of a theory involves as belief only that it is empirically adequate" and that if "the acceptance is at all strong, it is exhibited in the person's assumption of the role of explainer" (van Fraassen, 1980, p. 12). These assertions about the acceptance of T imply that we can invoke T to explain events even if we do not believe the truth of T.

Contra these philosophers, I claim that we cannot use T to explain events if we do not believe it; in other words, explainers using T to explain events are required to believe it (Park, 2014c, pp. 280–281). Antirealists do not believe T, and thus they cannot use it to explain events. If they use it to explain events, they run into one of the three problems: Moore's paradox, the problem of disconcerting questions, and the problem of deceptive speech acts. Let me expound these three problems one by one by using the law of gravity as an example.

According to the law of gravity, a baseball falls down due to gravitational force between it and the earth. Antirealists have to say, "A baseball falls down due to gravitational force" to explain why the baseball falls down in light of the law of gravity. They would have to add, however, "I don't believe a baseball falls down due to gravitational force" because they do not believe the law of gravity. If they say these two sentences consecutively, they are caught in Moore's paradox.

Moore's paradox arises from the assertion of any sentence that has the form, "P, but I don't believe p" (Moore, 1993). A Moorean sentence does not have persuasive force, and its assertion only puzzles the audience. Imagine that you are a police officer and that a person visits your police station. She states, "A mugger assaulted me, but I don't believe a mugger assaulted me." Her statement would only puzzle you, and it would not motivate you to take any action (Park, 2018c, p. 35). By similar reasoning, antirealists' speech act above would only puzzle their audience, and it would not be convincing to them. As a result, they would not be relieved of the puzzlement over why a baseball falls down.

Suppose now that explainers say, "A baseball falls down due to gravitational force." They do not say afterwards, "But I don't believe a baseball falls down due to gravitational force." Under this condition, they would not run into Moore's paradox. However, if their explainees know that the explainers are antirealists, the explainees might ask the explainers the following perplexing questions: "Do you believe what you just said? If you don't, why should I believe what you don't? How can you say to me what you don't believe? Do you expect me to believe what you don't?" (Park, 2019b, p. 155). It is murky how antirealists would respond to these disconcerting questions. This problem for antirealists might be called "the problem of disconcerting questions." It arises when antirealists say what they do not believe, but they do not add that they do not believe what they say.

Suppose now that explainers say, "A baseball falls down due to gravitational force." They do not say afterwards, "But I don't believe a baseball falls down due to gravitational force." Their explainees do not know that the explainers are antirealists. Under these conditions, antirealists would neither run into Moore's paradox nor into the problem of disconcerting questions. However, they would run into an ethical problem called "the problem of deceptive speech acts" (Park, 2015, pp. 226–228). Antirealists' speech acts mislead their explainees into believing what antirealists do not believe. For example, antirealists do not believe that a baseball falls down due to gravitational force. However, they do say, "A baseball falls down due to gravitational force." As a result of antirealists' speech act, their explainees might believe that a baseball falls down due to gravitational force, and antirealists might propagate beliefs to others that they do not have themselves.

Why is this speech act immoral? Suppose that you are an atheist, and that you yearn for an eternal life. You come across a cult leader. The cult leader says, "My god is so powerful that she can grant you an eternal life, provided that you worship her every day." You are persuaded by her preaching. As a result, you give up atheism and become her follower. You live a religious life for a few decades, just as she recommends. One day, she confesses to you that she has been an atheist all along. You would think that she deceived you and that her speech acts were immoral (Park, 2015, p. 227).

There is an important similarity between the cult leader and antirealists. They say what they do not believe, so they are merely pretending. The cult leader pretends to be a theist, and antirealists pretend to be realists. The cult leader speaks religious sentences although she does not believe them. She cannot speak the sentences without the intention to deceive her audience. Similarly, antirealists say, "A baseball falls down due to gravitational force," although they do not believe what they say. They cannot speak the sentence without the intention to deceive their explainees (Park, 2015, p. 227). Both the cult leader's speech act and antirealists' speech act are immoral in that they disregard others' desire to acquire truths and avoid errors.

In contrast to antirealists, realists do not run into Moore's paradox. Suppose that they say, "A baseball falls down due to gravitational force, and I believe a baseball falls down due to gravitational force." This sentence has the structure, "P, and I believe p." It is not a Moorean sentence. Hence, the assertion of it does not give rise to Moore's paradox, and it has the possibility of being convincing to realists' audience.

The problem of disconcerting questions does not arise for realists either. Suppose that realists say, "A baseball falls down due to gravitational force," and that their explainees ask realists, "Do you believe what you just said? If you don't, why should I believe what you don't? How can you say to me what you don't believe? Do you expect me to believe what you don't?" (Park, 2019b, p. 155). Realists would happily answer that they do believe what they say and that they do expect their explainees to believe what they believe.

How about the problem of deceptive speech acts? Suppose that realists say, "A baseball falls down due to gravitational force." When they say so, they are not pretending to believe the law of gravity; they are rather speaking in accordance with what they believe. Moreover, they do not have the intention of deceiving their explainees. Their speech act reflects what they believe. As a result, they are not liable to the charge that their speech act is deceptive.

Why are realists free, while antirealists are not, from the aforementioned three problems? Realists believe scientific theories. It follows that when they explain events in terms of scientific theories, what they say coincides with what they believe. By contrast, antirealists do not believe scientific theories. It follows that when they explain events in terms of scientific theories, what they say diverges from what they believe. This divergence is the source of the three problems for antirealists.

Let me turn to another disadvantage with antirealism. If we do not believe that an event occurs, we do not bother to explain why it occurs. For example, we do not believe that a unicorn flies away from a tiger. As a result, we do not bother to explain why a unicorn would fly away from a tiger. In current science, theoretical or observational events are invoked to explain theoretical events. For example, physicists say that "a proton has mass because it interacts with the Higgs field" (Park, 2019b, p. 155). They also say that minute gravitational waves are generated when there is the collapse of a building. If scientists were to embrace antirealism, they would not believe that the two explananda occur, and hence they would not bother to explain why they occur (Park, 2019b, p. 156).

I do not rule out the possibility, though, that antirealists can avail themselves of what I call "observational explanations." In an observational explanation, an observational event is explained in terms of another observational event. For example, antirealists can say that a rock falls off a cliff because John pushed it. The aforementioned three problems do not arise when they give such explanations. After all, what they say coincides with what they believe. However, observational explanations were even available for our ancient ancestors in the Stone Age, who did not have science.

What can we conclude from this discussion? An explanation is an important means of making scientific progress. If scientists choose antirealism over realism as their philosophical framework, they could and would give fewer explanations, and hence they could and would make less scientific progress. By contrast, if scientists choose realism over antirealism, they would give more explanations, and hence they would make more scientific progress (Park, 2019b, p. 157).

## 3.6 Prediction

Can antirealists use T to predict events? It appears that the correct answer is "Yes." It is the commonsensical view in philosophy of science that instrumentalists, who do not believe T, can use T to predict events. Contrary to this commonsensical view, I argue that antirealists cannot use T to predict events because of the aforementioned three problems.

Imagine that theoretical physicists have just developed T, that they make a prediction with the use of T, and that the prediction is not yet confirmed. Their predictees are experimental physicists whose job is to conduct experiments to determine whether predictions are true. They ask theoretical physicists to justify the prediction, pointing out that they should first be convinced of the prediction to some degree before they spend their time and energy on performing a relevant experiment. An interesting question arises. Can theoretical physicists, who are antirealists, provide T to meet this challenge? My answer is "No." Theoretical physicists do not believe T. Accordingly, they cannot provide it to support their prediction. How can they provide a premise that they do not believe to support a conclusion? If they provided T to justify their prediction, they would run into one of the aforementioned three problems. As a result, experimental physicists would not be motivated to ascertain the truth of theoretical physicists' prediction (Park, 2015, p. 225).

However, suppose that realists have used T to make true predictions before, and that antirealists know that T has made true predictions. Under this condition, antirealists would be entitled to use T to predict similar events (Park, 2015, p. 225). For example, realists used Isaac Newton's mechanics to predict the trajectory of a rocket to the moon, and antirealists know what realists did with Newton's mechanics. Under this condition, antirealists would be entitled to use Newton's mechanics to predict the trajectory of a rocket to Mars, although they do not believe Newton's mechanics. When challenged to justify their prediction, antirealists would appeal not to the theoretical content of Newton's mechanics but rather to its past performance. In short, once T made true predictions, antirealists can use it to make predictions.

I must emphasize, though, that only realists can be the initial users of T. They believe T, and thus they can provide the theoretical content of T to justify its first prediction. By contrast, antirealists do not believe T, and thus they cannot provide the theoretical content of T to justify its first prediction.

In addition, scientists predict not only observational events but also theoretical events. For example, physicists say that if two heavy rocks collide with each other, small gravitational waves will be generated and that "if the Higgs field were turned off, a *proton* would become massless" (Park, 2019b, p. 156). Such predictions might be called "theoretical predictions." Antirealists *would* not make theoretical predictions because they do not believe that such events occur. In addition, they *could* not make such predictions due to the aforementioned three problems.

By contrast, realists believe T, so they can use it to make such predictions. They run into none of the aforementioned three problems that plague antirealists. Prediction is an important means of making progress in science. Consequently, scientists would make more progress if they chose realism over antirealism as their philosophical framework for scientific practice. This pragmatic advantage of realism over antirealism is summarized in the slogan: "I believe more. Therefore, I progress more" (Park, 2019b, p. 158).

## 3.7 Two Epistemic Disadvantages

As noted in Sect. 3.1 of this chapter, antirealism has an epistemic advantage over realism, i.e., antirealism involves less epistemic risk than does realism. That being said, I expose two epistemic disadvantages with antirealism in this section.

Suppose that scientists embrace antirealism and that as a result, they do not believe their colleagues' theories. In such a case, it would be a double standard for them to believe their own theories, other things being equal. For example, Henri Poincaré (1905/1952, p. 160) and Ernst Mach (1911/2015, p. 17) advocated the classic PI, and they did not believe their fellow scientists' theories. Consequently, they ought not to have believed their own theories, whatever their theories might have been (Park, 2018d, p. 10).

Suppose now that scientists embrace realism and that as a result, they believe their colleagues' theories. In such a case, it would not be a double standard, ceteris paribus, for them to believe their own theories. They treat their own theories and others' theories equally, and they abide by the epistemic imperative stated in Sect. 3.2 of this chapter. They will the maxim, "Believe a successful scientific theory," to become a universal one, and they believe their own successful theories in accordance with the maxim.

Suppose that *philosophers* embrace antirealism and that they do not believe scientific theories. In such a case, it would be unfair for them to believe their own philosophical theories, other things being equal. For example, van Fraassen does not believe scientific theories. Consequently, he should refrain from believing his own positive philosophical theories, which include the contextual theory, according to which "An explanation is an answer to a why-question" (van Fraassen, 1980, p. 134) and "among the scientifically relevant factors, context determines explanatorily relevant ones" (van Fraassen, 1980, p. 126).

Suppose now that philosophers embrace realism and that as a result, they believe scientific theories. In such a case, it would not be a double standard, ceteris paribus, for them to believe their own philosophical theories. Michael Friedman, for example, believes scientific theories (Friedman, 1981, p. 7) and advocates the unification theory of scientific explanation (Friedman, 1974). He can will the maxim, "Believe a successful theory, whether scientific or philosophical," to become a universal maxim, and he can believe the unification theory in accordance with the maxim.

The foregoing discussion shows that antirealism has an epistemic disadvantage: if you embrace it, you cannot believe your own theories, whether you are a scientist or a philosopher (Park, 2019c, p. 92). Realism has an epistemic advantage: if you embrace it, you can believe your own theories, whether you are a scientist or a philosopher. Consequently, realism is the right epistemic policy for *active* cognitive agents, i.e., those who generate theories. By contrast, antirealism is the right epistemic policy for *passive* cognitive agents, i.e., those who take an attitude toward others' theories.

It is rare to see altruistic moral agents, i.e., those who put others' interests above their own. We have the psychological inclination to be more lenient with ourselves and to be stricter with others concerning moral matters, other things being equal. Immanuel Kant observes that this psychological inclination is the source of immoral actions. This observation motivates him to come up with the categorical imperative: "Act only according to that maxim through which you can at the same time will that it should become a universal law" (Kant, 1785/2011, p. 71). He thinks that abiding by the categorical imperative is a way to bring about justice in our moral lives.

Similarly, it is rare to see altruistic epistemic agents, i.e., those who disbelieve their own theories and believe others' theories, other things being equal. We have the psychological inclination, ceteris paribus, to believe our own theories and to disbelieve our epistemic colleagues' theories. I contend that this psychological inclination is a source of unfair epistemic actions and that abiding by the epistemic imperative stated in Sect. 3.2 of this chapter is a way to bring about justice in our epistemic lives.

Let me elucidate an implication of disbelieving your own theories. Your attitude toward your own theory influences your epistemic colleagues' attitudes toward it. If you believe it, they will be more inclined to believe it. If you disbelieve it, they will be less inclined to believe it. Imagine, for example, that Antoine Lavoisier puts forward the oxygen theory at a scientific conference. If he declares at the end of his presentation that he disbelieves it, his audience will be less inclined to believe it as well (Park, 2018c, p. 35). Imagine also that van Fraassen presents his contextual theory at a philosophy conference. If he declares at the end of his presentation that he disbelieves it, his audience will be less inclined to believe it (Park, 2019d, p. 149).

Park (2019c, p. 91) argues that it is unfair for van Fraassen to believe the contextual theory and not to believe scientific theories. Playing devil's advocate, Healey (2019b, pp. 26–27) replies that van Fraassen does not believe his theory. I counter-reply that van Fraassen speaks as if he believes it (Park, 2019d, p. 148). Van Fraassen states, for instance, "An explanation is an answer to a why-question" (van Fraassen, 1980, p. 134) and "among the scientifically relevant factors, context determines explanatorily relevant ones" (van Fraassen, 1980, p. 126). Note that he speaks as if he believes his theory. If he does not believe it, we can conclude that his sentences mislead his readers into believing what he does not believe. If he does not want to mislead his readers, he should put a caveat in his book, "I do not believe the contextual theory." If he does so, however, his epistemic colleagues will be less motivated to read his book.

By contrast, realism does not have this epistemic disadvantage. Realists believe their own theories, whether scientific or philosophical. Their positive attitude toward their own theories makes their audience more inclined to believe their theories. For example, Friedman believes scientific theories and the unification theory of scientific explanation. His positive attitude toward the unification theory makes his audience more inclined to believe it, and he does not need to put in his book the caveat: "I do not believe the unification theory."

In sum, antirealism has two epistemic disadvantages: if you embrace it, you ought not to believe your own positive theories, and you are less likely to spread them to your epistemic colleagues. By contrast, realism has two epistemic advantages: if you embrace it, you can believe your own positive theories, and you are more likely to spread them to your epistemic colleagues. Again, realism is the right epistemic policy for active cognitive agents, and antirealism is the right epistemic policy for passive cognitive agents.

## 3.8 Epistemic Reciprocalism

This section raises objections to antirealism from the perspective of the position called "epistemic reciprocalism." Epistemic reciprocalism asserts that "we ought to treat our epistemic colleagues, as they treat their epistemic colleagues, *ceteris paribus*" (Park, 2017c, p. 57). Epistemic reciprocalism is developed under the framework of social epistemology (Goldman, 1999), according to which epistemic agents are not isolated from one another but rather interact with one another.

How would epistemic reciprocalists respond to realists and antirealists? Imagine that realists and antirealists attend the same academic conference. Realists put forward positive theories, whatever they might be. They show the audience what phenomena their theories can explain and predict. So do antirealists. Realists' theories and antirealists' theories are confirmed more or less to the same degree. To antirealists' surprise, however, the audience believes realists' theories and disbelieves antirealists' theories. Antirealists ask the audience to explain why they disbelieve their theories. The audience responds that they are epistemic reciprocalists! They add that if antirealists want others to believe their theories, they should believe others' theories. After all, "There is no reason for thinking that the Golden Rule ranges over moral matters, but not over epistemic matters" (Park, 2018e, pp. 77–78).

Suppose that some of your epistemic colleagues are epistemic egoists (Park, 2017c, p. 59). They believe their own theories but disbelieve others' theories, despite the fact that their own theories and others' theories are more or less of the same epistemic status. You complain to them that it is a double standard for them to believe their own theories but to disbelieve others' theories. To your disappointment, they ignore your complaint and stick to epistemic egoism. How should you respond to them? An answer to this question can be found in epistemic reciprocalism.

Epistemic reciprocalists believe their own theories but disbelieve epistemic egoists' theories. As a result, epistemic egoists fail to propagate their theories to epistemic reciprocalists. To escape from this quagmire, epistemic egoists might believe epistemic reciprocalists' theories, expecting that epistemic reciprocalists will, in return, believe their theories. As expected, epistemic reciprocalists would believe former epistemic egoists' theories. This imaginary interaction between epistemic egoists and reciprocalists illustrates how epistemic reciprocalism works in our epistemic lives. The spirit of epistemic reciprocalism is expressed in the slogan: "Believe me. I'll believe you" (Park, 2017c, p. 64).

Admittedly, we are interested in avoiding false beliefs. For this reason, we carefully assess our epistemic colleagues' arguments for their theories before we believe them. However, we are also interested in propagating "to others our own theories which we are confident about" (Park, 2017c, p. 58). For this reason, we construct strong arguments for our own theories. We usually feel excited and thrilled when our epistemic colleagues believe our theories. By contrast, we usually feel depressed and frustrated when they disbelieve our theories. Antirealists are more likely to feel depressed and frustrated than realists when they interact with epistemic reciprocalists.

Epistemic reciprocalism can be used against van Fraassen's contextual theory (Park, 2017c, pp. 61–62). Van Fraassen employs inference to the best explanation (IBE) to justify the contextual theory, as noted in Sect. 1.5 of Chap. 1. Epistemic reciprocalists treat the contextual theory in the same way van Fraassen treats scientific theories, which means that they do not believe it. They also point out that van Fraassen should not believe it and that it would be a double standard for him to believe it.

This epistemic disadvantage of antirealism can incur pragmatic disadvantages. Suppose that the contextual theory is so virtuous that contextualists believe that they deserve a scholarly award and that they submit an application to an award committee. To their disappointment, the committee members reject the application. They have embraced epistemic reciprocalism. As a result, they refuse to believe the contextual theory. Moreover, the committee has a regulation, according to which the committee members should believe applicants' theories to grant an award to them.

The committee members reject contextualists' application pursuant to this regulation (Park, 2019c, p. 96).

Contextualists might challenge the committee members to explain why they refuse to believe the contextual theory. The committee members would remind contextualists that van Fraassen puts forward the argument from a bad lot to attack IBE, as we noted in Sect. 1.5 of Chap. 1. The argument from a bad lot can just as well be directed at the contextual theory. There are a collection of conceived theories of explanation and a collection of unconceived theories of explanation. Before van Fraassen believes the contextual theory, he should establish that a true theory of explanation is more likely to exist in the former collection than in the latter collection. So would the committee members explain to contextualists.

Realists do not suffer from such pragmatic disadvantages, but rather enjoy pragmatic advantages. Suppose that realists have a positive philosophical theory and that they believe it on the grounds that it is successful. They think that they deserve a scholarly award, and then they submit an application to an award committee. To their joy, the committee members grant them the award. They have embraced epistemic reciprocalism. As a result, they believe realists' theory. This example illustrates that our epistemic colleagues' epistemic endorsement might come with their pragmatic endorsement. It is natural that we have the inclination to feel excited and thrilled when our epistemic colleagues believe our theories.

Van Fraassen might retort that he believes that scientific theories are EA, although he does not believe that they are true, so epistemic reciprocalists should believe that the contextual theory is EA. There is a considerable difference between believing the truth and believing the empirical adequacy of the contextual theory. To believe the truth of the contextual theory is to believe that "An explanation is an answer to a why-question" (van Fraassen, 1980, p. 134) and "among the scientifically relevant factors, context determines explanatorily relevant ones" (van Fraassen, 1980, p. 126). By contrast, to believe the empirical adequacy of the contextual theory is to believe that "explananda of the contextual theory, such as rejections and asymmetries, occur in science" (Park, 2019c, p. 94).

If contextualists believe that scientists' theories are EA, epistemic reciprocalists would, in response, believe that the contextual theory is EA. However, the award committee might still refuse to grant contextualists the award, maintaining that giving the award to applicants requires that the committee members believe that applicants' theories are true. It is rational for an award committee to enact and enforce the regulation (Park, 2019c, pp. 95–96).

The spirit of epistemic reciprocalism permeates many corners of this book. It can be found in the grand PI against the six first-order PIs in Sect. 2.4 of Chap. 2, in the PIs over pessimists and antirealists Sects. 3.2 and 3.3 of this chapter, and in the six new arguments for realism that will be unpacked in Chap. 4. All of these arguments against antirealism and for realism take advantage of the very ideas that antirealists have come up with to attack realism, and hence they all accord with epistemic reciprocalism.

## 3.9 Empiricism and the Problem of Unobserved Anomalies

In Sect. 1.9 of Chap. 1, I gave the name "semantic empiricism" to the position that T, a successful theory, is EA and "epistemic empiricism" to the position that we can justifiably believe that T is EA. I also argued that semantic empiricism provides a circular explanation of the success of T. This section aims to expose another problem of empiricism.

Wray (2008, 2010) makes the following two conflicting claims. On the one hand, he says that "all that scientists can claim to know about their background theories is that they save the phenomena" (Wray, 2008, p. 321). This claim can be interpreted as advancing epistemic empiricism. On the other hand, Wray endorses Stanford's PUA, saying "that future developments in a field will reveal additional competing theories that are also able to account for the data, theories, though, that are now *unconceived* alternatives" (Wray, 2010, p. 371). The PUA clashes with empiricism. The PUA implies that the downfall of current theories can be expected from that of their predecessors. Past theories were overthrown due to their anomalies. Galileo, for example, observed the surface of the moon with his telescope. His observations undercut the Ptolemaic theory. Such anomalies revealed that past theories were not EA. We can inductively infer that current theories also have anomalies, and hence they are not EA either. It follows that we cannot justifiably believe that T is EA.

Wray might retort that many current theories do not have anomalies and thus that such theories are EA. For example, special relativity has passed every test that it has faced since it was born in 1905. My response to this observation of current theories is to raise the problem of unobserved anomalies, according to which "since past scientists could not observe the anomalies that caused the replacement of their theories with present theories, present scientists cannot observe the anomalies either that will cause the replacement of present theories with future theories" (Park, 2018d, p. 5). For example, Ptolemy could not observe the phases of Venus, but Galileo observed them. Similarly, current scientists cannot observe the anomalies of special relativity, but future scientists will be able to. The notion of unobserved anomalies dampens empiricists' aspirations for empirical adequacy, if the notion of unconceived alternatives dampens realists' aspiration for truth.

In addition, empiricists, who believe that T is EA, cannot use T to explain and predict events because of the aforementioned three problems. Consider, again, the law of gravity. The belief that the law of gravity is EA does not imply that gravitational force exists (Park, 2017c, p. 60). Consequently, empiricists say more than what their beliefs allow them to say when they say, "A baseball falls down due to gravitational force." The divergence between what they believe and what they say gives rise to the three problems. For example, they run into Moore's paradox when they say, "A baseball falls down due to gravitational force, but I don't believe that a baseball falls down due to gravitational force."

## 3.10 Constructive Empiricism

Does the problem of unobserved anomalies spell trouble for constructive empiricism (van Fraassen, 1980)? It would, if constructive empiricism asserts that T is EA. However, it is a thorny issue whether constructive empiricism entails that T is EA. This section attempts to resolve this issue with the view toward criticizing constructive empiricism, which is widely regarded as a form of antirealism.

Constructive empiricism is composed of the following two theses: "Science aims to give us theories which are empirically adequate," and "acceptance of a theory involves as belief only that it is empirically adequate" (van Fraassen, 1980, p. 12). Let me call the first thesis "axiological empiricism" and the second thesis "attributive empiricism." It is clear that axiological empiricism does not entail that T is EA. After all, there is a considerable difference between the thesis that we have EA theories and the thesis that we seek EA theories.

How about attributive empiricism? To answer this question, we should be clear about van Fraassen's technical notion of acceptance. For him, the acceptance of a theory involves "a commitment to confront any future phenomena by means of the conceptual resources of this theory" (van Fraassen, 1980, p. 12). Suppose, for example, that geologists accept plate tectonics and then confront the world by means of plate tectonics, i.e., they predict and explain events in terms of plate tectonics. According to attributive empiricism, when geologists do these things, they believe that plate tectonics is EA, but they do not believe that plate tectonics is true. As this example illustrates, attributive empiricism attributes the belief that T is EA to accepters of T, i.e., those who commit to confront the world in terms of T (Park, 2020b, p. 484). Consequently, this position is named "attributive empiricism."

Does attributive empiricism assert that T is EA? There is a reason for thinking that it does not. It is not concerned with whether T is EA, but rather with whether the acceptance of T involves the belief that T is EA. Instead of asserting that T is EA, it asserts that accepters of T believe that T is EA, i.e., it attributes the belief that T is EA to accepters of T. It is one thing to attribute a belief to the accepters of T; it is entirely another for attributors to hold the belief themselves. To put it differently, attributors might not hold a belief when attributing it to the accepters of T. It appears, therefore, that attributive empiricism does not entail that T is EA.

However, there is a reason for thinking that attributive empiricism asserts that T is EA. Attributive empiricists have to accept T. If they do not, they cannot confront the world in terms of it, and their daily lives would be unpalatable and savage. Just imagine what would happen to your daily life if you did not avail yourself of scientific predictions. Now, when attributive empiricists accept T, they should attribute the belief that T is EA to themselves. If they do not, they would cease to be attributive empiricists (Park, 2019d, p. 150). Therefore, attributive empiricists believe that T is EA.

Attributive empiricists might reply that they merely accept T for the sake of convenience in their daily lives, which implies that they attribute to themselves the belief that T is EA, but they do not hold the belief. It follows that they are free from the problem of unobserved anomalies. In my view, however, it is a symptom of irrationality for attributive empiricists to attribute to themselves the belief that T is EA when they do not hold it. To use an analogy, it is a symptom of irrationality for atheists to attribute to themselves the belief that God exists, when they do not hold the belief. In general, if you do not hold a belief, you know that you do not hold it; if you hold a belief, you know that you do hold it. If you do not hold a belief, you cannot persuade yourself that you do hold it, and hence you cannot attribute it to yourself.

Moreover, the problem of unobserved anomalies makes it more problematic for attributive empiricists to attribute to themselves the belief that T is EA. Suppose that they take the problem of unobserved anomalies seriously and that as a result they take it to be false that T is EA. They nevertheless accept T for the convenience of their daily lives, which implies that they attribute to themselves the belief that T is EA and that they take the belief to be false. It is a bizarre cognitive state of affairs that they attribute to themselves a belief that they take to be false.

Van Fraassen has never stated whether he is committed to the empirical adequacy of T, so it is not surprising that "It is a perennial issue what van Fraassen's position is" (Park, 2020c, p. 38). However, it is not important to determine whether he is committed to the empirical adequacy of T, since we can construct the following dilemma against constructive empiricism. Constructive empiricism either asserts or does not assert that T is EA. If it does, it falls prey to the problem of unobserved anomalies. If it does not, constructive empiricists ought not to believe that their positive theories are EA, and epistemic reciprocalists will not believe that their positive theories are EA (Park, 2019d, pp. 149–151). It will be interesting to see how constructive empiricists would tackle this dilemma. In addition, as we have seen in Sect. 3.8 of this chapter, constructive empiricism is vulnerable to the aforementioned three problems, whether or not it is committed to empiricism that T is EA.

What do commentators say about constructive empiricism? Some affirm while others deny that it asserts that T is EA. For example, Stathis Psillos (1997, p. 370) observes that it is an ampliative inference that since T is successful, it is EA. Wray replies that "the constructive empiricist is not committed to claiming that our best theories are in fact empirically adequate" (Wray, 2012, p. 378). Park (2019c, pp. 92–93) asserts that constructive empiricism is committed to the empirical adequacy of T and that constructive empiricism has epistemic and pragmatic disadvantages as compared to realism. Playing devil's advocate, Healey replies that constructive empiricists believe neither that T "is true nor that it is empirically adequate" (Healey, 2019b, p. 25). As these dialectics illustrate, when critics criticize the position that T is EA, defenders of constructive empiricism tend to say, "You misrepresented van Fraassen's position" (Park, 2019d, p. 153).

Mizrahi (2018) makes a similar observation in the context of the debate between opponents and proponents of Thomas Kuhn's philosophy of science. Mizrahi says, "Whenever the work of an influential philosopher is criticized, a common move made by those who seek to defend the influential philosopher's work is to claim that his or her ideas have been misconstrued" (Mizrahi, 2018, p. 19). Mizrahi points out that there is a problem with this common move. If it is made too often, critics might doubt that "the influential philosopher's ideas are worthy of attention and/or acceptance" (Mizrahi, 2018, p. 19). Let me add that "There is no difference between an elusive idea and an empty one, or at least we should treat an elusive idea as if it were empty" (Park, 2020b, pp. 492–493). The justification for this treatment is that "We can learn nothing from an elusive idea" (Park, 2020b, p. 492).

Whether or not constructive empiricism asserts that T is EA, it implies that accepters of T run into the three problems: Moore's paradox, the problem of disconcerting questions, and the problem of deceptive speech acts. Suppose that you accept the theory of gravity and then say to your explainees, "A baseball falls down due to gravitational force." According to constructive empiricism, when you say so, you do not believe that the theory of gravity is true, and you rather believe that it is EA. So what? You have to add, "But I don't believe that a baseball falls down due to gravitational force," thereby running into Moore's paradox. What if you do not add the second sentence? In case your explainees know that you believe that the theory of gravity is EA, they will ask you the aforementioned disconcerting questions. In case your explainees do not know that you believe that the theory of gravity is EA, they would assume that you believe that the theory of gravity is true, so they would not ask the disconcerting questions. However, your speech act is immoral because it misleads your audience into believing what you do not believe (Park, 2015, pp. 226–228). You disregard your explainees' desire to acquire truths and avoid errors. In short, constructive empiricists inadvertently attribute the three problems to scientists, the prominent accepters of T. If constructive empiricists accept T themselves, they inadvertently attribute the three problems to themselves.

## 3.11 Gap Antirealists

I have raised several objections to antirealism in this chapter. How would antirealists try to defend antirealism against my objections? They might play what I call "the gap card" (Park, 2014d, p. 269). To play the gap card is to say that there is a gap between what one says and what one believes, so a criticism against what one says does not refute what one believes. Let me call "gap antirealists" those antirealists who attempt to nullify the objections to antirealism by saying that there is a gap between what they say and what they believe.

Are there gap antirealists in the literature? I have argued that it is self-refuting for Kuhn (1962/1970) to appeal to evolutionary theory in order to defend his pessimistic account of scientific development, according to which new theories are no more truthlike than previous ones (Park, 2017d, pp. 324–325). Amanda Bryant retorts, "Kuhn could adopt the language of a paradigm (for the purposes of drawing an analogy, no less!) without committing to the literal truth of that paradigm" (Byrant, 2018, p. 3). In other words, Kuhn can use evolutionary language although he does not believe the theory of evolution. For example, Kuhn can say that organisms are "products of a process that moved steadily *from* primitive beginnings but *toward* no goal" (Kuhn, 1962/1970, p. 172). Byrant's retort implies that there can be a gap between what Kuhn says and what he believes, so my criticism might not refute what Kuhn believes, although it might refute what he says. Obviously, she thinks that gap antirealists are feasible cognitive agents.

How do I respond to gap antirealism? I point out that it has an epistemic disadvantage. Gap antirealists' positive theories would be taken less seriously by their cognitive colleagues. Imagine that gap antirealists publish a book in which they expound positive theories, whatever they might be. There should be a caveat on the front cover that says, "There is a gap between what we wrote and what we believe. You will never be able to refute what we believe, no matter how brilliant and sharp your criticisms might be." Such a caveat would repel many readers, since they would take what is written in the book less seriously.

What if gap antirealists refuse to put the caveat on the front cover? My response is to say, "If we take in our hand the gap antirealists' book, let us ask, 'Does it have the caveat that the authors do not believe what they wrote in the book?' No. Commit it then to the flames, for it misleads their readers" (Park, 2019d, pp. 148–149). These sentences echo David Hume's famous sentences: "If we take in our hand any volume—of divinity or school metaphysics, for instance—let us ask, *Does it contain any abstract reasoning concerning quantity or number?* No. *Does it contain any experimental reasoning concerning matter of fact and existence?* No. Commit it then to the flames, for it can contain nothing but sophistry and illusion" (Hume, 1777/1977, p. 114).

A pragmatic disadvantage with gap antirealism can be brought to light with the use of a more realistic example. Imagine that gap antirealists are professors who teach philosophy courses to undergraduate students. After the final exam, some of the students complain about their grades. The students argue that there was a gap between what they wrote and what they thought, so they deserve better grades. They have embraced epistemic reciprocalism! Gap antirealists have no choice but to raise their grades. They cannot say to their students, "Professors go by what students wrote, not what they thought." After all, to say so goes against their philosophical position. As a result, they end up treating unfairly those students who do not complain about their grades.

## 3.12 Summary

We can construct the PI over antirealist explanations on the basis of the intrinsic flaws with several antirealist explanations, the PI over pessimists on the basis of the historical fact that recent past theories have been mostly stable, and the PI over antirealists on the basis of the historical fact that antirealists have made many philosophical mistakes. Antirealists cannot explain events in terms of T due to Moore's paradox, the problem of disconcerting questions, and the problem of deceptive speech acts. Antirealists will not and cannot predict theoretical events because they do not believe that such events occur. They cannot believe their own positive theories because they would have to treat their own theories and others' theories equally.

Epistemic reciprocalists do not believe antirealists' positive theories because antirealists do not believe others' positive theories. The problem of unobserved anomalies refutes the position that T is EA. Not believing that T is EA comes with epistemic and pragmatic disadvantages. Constructive empiricism attributes to accepters of T the three problems: Moore's paradox, the problem of disconcerting questions, and the problem of deceptive speech acts. Playing the gap card comes with epistemic and pragmatic disadvantages.

## References


Brown, J. (1994). *Smoke and mirrors: How science reflects reality*. Routledge.

Byrant, A. (2018). Each Kuhn mutually incommensurable. *Social Epistemology Review and Reply Collective, 7*(6), 1–7.

Dawes, G. (2013). Belief is not the issue: A defence of inference to the best explanation. *Ratio: An International Journal of Analytic Philosophy, 26*(1), 62–78.

Dellsén, F. (2017). Understanding without justification or belief. *Ratio: An International Journal of Analytic Philosophy, 30*(3), 239–254.

Fahrbach, L. (2011). How the growth of science ends theory change. *Synthese, 180*(2), 139–155.

Fine, A. (1986). Unnatural attitudes: Realist and instrumentalist attachments to science. *Mind, 95*(378), 149–179.

Fine, A. (1991). Piecemeal realism. *Philosophical Studies, 61*(1–2), 79–96.

Friedman, M. (1974). Explanation and scientific understanding. *The Journal of Philosophy, 71*(1), 5–19.

Friedman, M. (1981). Theoretical explanation. In R. Healey (Ed.), *Reduction, time & reality: Studies in the philosophy of natural science* (pp. 1–16). Cambridge University Press.

Goldman, A. (1999). *Knowledge in a social world*. Oxford University Press.

Healey, R. (2019a). Four points in response to Seungbae Park. *Social Epistemology Review and Reply Collective, 8*(12), 43–45.

Healey, R. (2019b). The aims of reliable knowledge: Reply to Seungbae Park. *Social Epistemology Review and Reply Collective, 8*(9), 25–30.

Hume, D. (1777/1977). *An enquiry concerning human understanding*. Hackett Publishing Company.

Kant, I. (1785/2011). In J. Timmermann (Ed.), *Groundwork of the metaphysics of morals*. Cambridge University Press.

Kuhn, T. (1962/1970). *The structure of scientific revolutions*. University of Chicago Press.

Kukla, A. (1998). *Studies in scientific realism*. Oxford University Press.

Ladyman, J., Douven, I., Horsten, L., & van Fraassen, B. (1997). A defense of van Fraassen's critique of abductive inference: Reply to Psillos. *The Philosophical Quarterly, 47*(188), 305–321.

Laudan, L. (1984). Explaining the success of science: Beyond epistemic realism and relativism. In J. Cushing, C. Delaney, & G. Gutting (Eds.), *Science and reality* (pp. 83–105). University of Notre Dame Press.

Lyons, T. (2003). Explaining the success of a scientific theory. *Philosophy of Science, 70*(5), 891–901.

Mach, E. (1911/2015). *History and root of the principle of the conservation of energy*. Trans. P. Jourdain. Open Court Publishing Company.

Mizrahi, M. (2013). The pessimistic induction: A bad argument gone too far. *Synthese, 190*(15), 3209–3226.

Mizrahi, M. (2016). The history of science as a graveyard of theories: A philosophers' myth. *International Studies in Philosophy of Science, 30*(3), 263–278.

Mizrahi, M. (2018). The (lack of) evidence for the Kuhnian image of science. *Social Epistemology Review and Reply Collective, 7*(7), 19–24.

Moore, G. (1993). Moore's paradox. In T. Baldwin (Ed.), *G.E. Moore: Selected writings* (pp. 207–212). Routledge.

Musgrave, A. (1988). The ultimate argument for scientific realism. In R. Nola (Ed.), *Relativism and realism in science* (pp. 229–252). Kluwer Academic Publishers.

Park, S. (2001). *Scientific realism vs. scientific antirealism*. Ph.D. Dissertation. University of Arizona.

Park, S. (2011). A confutation of the pessimistic induction. *Journal for General Philosophy of Science, 42*(1), 75–84.

Park, S. (2014a). A pessimistic induction against scientific antirealism. *Organon F, 21*(1), 3–21.

Park, S. (2014b). Approximate truth vs. empirical adequacy. *Epistemologia, 37*(1), 106–118.

Park, S. (2014c). The doxastic requirement of scientific explanation and understanding. *PRO, 13*(2), 279–290.

Park, S. (2014d). On the evolutionary defense of scientific antirealism. *Axiomathes, 24*(2), 263–273.

Park, S. (2015). Accepting our best scientific theories. *Filosofija. Sociologija, 26*(3), 218–227.

Park, S. (2017a). Why should we be pessimistic about antirealists and pessimists? *Foundations of Science, 22*(3), 613–625.

Park, S. (2017b). Scientific antirealists have set fire to their own houses. *PRO, 16*(1), 23–37.

Park, S. (2017c). Defense of epistemic reciprocalism. *Filosofija. Sociologija, 28*(1), 56–64.

Park, S. (2017d). Problems with using evolutionary theory in philosophy. *Axiomathes, 27*(3), 321–332.

Park, S. (2018a). In defense of the epistemic imperative. *Axiomathes, 28*(4), 435–446.

Park, S. (2018b). Philosophers and scientists are social epistemic agents. *Social Epistemology Review and Reply Collective, 7*(6), 31–43.

Park, S. (2018c). The problem of unobserved anomalies. *Filosofija. Sociologija, 29*(1), 4–12.

Park, S. (2018d). The pessimistic induction and the Golden rule. *PRO, 93*, 70–80.

Park, S. (2019a). The absolute and relative pessimistic inductions. *PRO, 95*, 94–104.

Park, S. (2019b). Should scientists embrace scientific realism or antirealism? *Philosophical Forum, 50*(1), 147–158.

Park, S. (2019c). The disastrous implications of the 'English' view of rationality in a social world. *Social Epistemology, 33*(1), 88–99.

Park, S. (2019d). Constructive empiricism in a social world: Reply to Richard Healey. *Social Epistemology Review and Reply Collective, 8*(10), 146–154.

Park, S. (2020a). How to overcome antirealist objections to scientific realism. *Axiomathes, 30*(1), 1–12.

Park, S. (2020b). Formulational vs. epistemological debates concerning scientific realism. *Dialogue: Canadian Philosophical Review, 59*(3), 479–496.

Park, S. (2020c). Replies to Healey's comments regarding van Fraassen's positions. *Social Epistemology Review and Reply Collective, 9*(1), 38–47.

Poincaré, H. (1905/1952). *Science and hypothesis*. Dover.

Psillos, S. (1997). How not to defend constructive empiricism: A rejoinder. *The Philosophical Quarterly, 47*(188), 369–372.

Psillos, S. (1999). *Scientific realism: How science tracks truth*. Routledge.

Stanford, K. (2000). An antirealist explanation of the success science. *Philosophy of Science, 67*(2), 266–284.

Stanford, K. (2015). Catastrophism, uniformitarianism, and a scientific realism debate that makes a difference. *Philosophy of Science, 82*(5), 867–878.

van Fraassen, B. (1980). *The scientific image*. Oxford University Press.

van Fraassen, B. (1989). *Laws and symmetry*. Oxford University Press.

Winther, R. (2009). A dialogue. *Metascience, 18*, 370–379.

Wray, B. (2008). The argument from underconsideration as grounds for anti-realism: A defence. *International Studies in the Philosophy of Science, 22*(3), 317–326.
Wray, B. (2010). Selection and predictive success. *Erkenntnis, 72*(3), 365–377.
Wray, B. (2012). Epistemic privilege and the success of science. *Noûs, 46*(3), 375–385.
Wray, B. (2013). The pessimistic induction and the exponential growth of science reassessed. *Synthese, 190*(18), 4321–4330.
Wray, B. (2015). Pessimistic inductions: Four varieties. *International Studies in the Philosophy of Science, 29*(1), 61–73.

# Chapter 4
# Six New Arguments for Scientific Realism

**Abstract** I introduce the following six new arguments for realism, all of which radically differ from the no-miracles argument. The optimistic induction over realists holds that since the realists of the early twentieth century were right to believe the theories of their time, their present and future philosophical offspring are also right to believe the theories of their times. The argument from undiscovered evidence holds that because T has been supported by previously undiscovered evidence, it will continue to be supported by hitherto undiscovered evidence. The argument from neighboring theories holds that since T has been reinforced by previously unconceived neighboring theories, it will continue to be reinforced by hitherto unconceived neighboring theories. The counter-induction for realism asserts that since earlier theories were ousted, their present successors will not be ousted. The English argument for realism holds that under the English notion of rationality, it is rational to believe T, although it is also rational not to believe T. These five new arguments for realism entitle us to construct a second-order argument for scientific realism that I call "the grand optimistic induction." It holds that since realism has been supported by the five previously unconceived arguments, it will continue to be supported by hitherto unconceived arguments.

**Keywords** Neighboring theories · Undiscovered evidence · Counter-induction · English argument · Optimistic induction

## 4.1 Introduction

The no-miracles argument (NMA) has been the main argument for realism ever since Hilary Putnam (1975) constructed it, as noted in Sect. 1.1 of Chap. 1. This chapter aims to introduce six new arguments for realism. They each have the following two characteristics. First, they are all profoundly distinct from the NMA. Second, they utilize the very ideas that antirealists have developed to attack realism. It follows that they are no worse off than antirealists' arguments against realism and that


S. Park, *Embracing Scientific Realism*, Synthese Library 445,
https://doi.org/10.1007/978-3-030-87813-9_4

any future criticisms against them will backfire against antirealists' arguments against realism.

The structure of this chapter is as follows. In Sect. 4.2, I elucidate the optimistic induction over realists (Park, 2019a), which claims that just as realists from the recent past were right to believe the scientific theories of their time, so their present and future philosophical offspring are also right to believe the scientific theories of their times. In Sect. 4.3, I unpack the argument from undiscovered evidence (Park, 2018a), which asserts that since T has been supported by previously undiscovered evidence, it will continue to be supported by hitherto undiscovered evidence. In Sect. 4.4, I expound the argument from neighboring theories (Park, 2019b), which holds that since T has been bolstered by previously unconceived theories in the neighboring fields of science, it will continue to be bolstered by hitherto unconceived theories in the neighboring fields of science. In Sect. 4.5, I unfold the counter-induction for realism (Park, 2018b), which says that because earlier theories were discarded, current ones will be retained. In Sect. 4.6, I explicate the English argument for realism (Park, 2020), which maintains that under the English notion of rationality, it is rational to believe T, although it is also rational not to believe T. In Sect. 4.7, I construct the grand optimistic induction, according to which since realism has been reinforced by the five previously unconceived arguments, it will continue to be bolstered by hitherto unconceived arguments.

## 4.2 The Optimistic Induction Over Realists

Recent past theories, such as special relativity, have remained unfalsified for about one hundred years, as we noted in Subsection 2.4.1 of Chap. 2. This historical fact entitles us to construct the optimistic induction for realism, according to which since recent past theories have been accepted, they "will remain accepted" (Park, 2011, p. 80). This argument can be given a little twist as follows. Imagine that the optimists of the early twentieth century believed the theories of their time on the grounds that they were more successful than their predecessors. We can tell from our vantage point of the twenty-first century that they were right to believe the theories of their time. The optimists of the twenty-first century also believe the theories of their time on the grounds that they are more successful than their predecessors. Our descendants of the twenty-second century will be able to see from their vantage point that the optimists of the twenty-first were right to believe the theories of their time. This induction is called "the optimistic induction over realists" (Park, 2019a, p. 98).

The optimistic induction over realists parallels Brad Wray's (2013) and Thomas Nickles's (2017) pessimistic induction (PI) over realists that was discussed in Subsection 2.4.4 of Chap. 2. Wray and Nickles imagine that distant past realists believed their theories on the grounds that their theories were better than their predecessors. We can tell from the vantage point of the twenty-first century that their beliefs were false. Present realists believe present theories on the grounds that they are better than their precursors. However, "what looks new to us will not look so new to our offspring" (Wray, 2013, p. 4327). Wray and Nickles construct the PI over realists to overcome the objection to the classic PI that present theories are better than their precursors. Just as Wray and Nickles run the PI over realists, so I run the optimistic induction over realists. The optimistic induction over realists is better than the PI over realists because current theories are more similar to recent past ones than to distant past ones.

Realism, which is defended by the optimistic induction over realists, substantially differs from selective realism. According to selective realism, the stable constituents of a theory are believable, whereas the unstable constituents of a theory are unbelievable (Worrall, 1989; Psillos, 1999, 2018; Chakravartty, 2008; Vickers, 2017, 2018). When a scientific revolution occurs, some parts of a theory are retained, while others are abandoned. In addition, the former are deployed to generate predictions and explanations, while the latter merely serve as metaphysical frameworks. Consequently, we can justifiably believe the former, but not the latter. Selective realism thus sketched asserts, as does pessimism, that current theories will be ousted (Park, 2017a, p. 99, 2019a, p. 94; Stanford, 2018, p. 79). By contrast, realism asserts that they will not; rather, they will go through at most minor modifications, and they are overall (approximately) true.

Kyle Stanford's (2009, p. 385) objection to selective realism does not apply to realism. He points out that there should be a criterion for differentiating between the stable and unstable constituents of current theories. Without that criterion, we would not know which constituents of current theories are believable and which are unbelievable. Selective realism is open to this criticism because it asserts that current theories will be ousted. By contrast, realism is not open to this criticism because it does not make the assertion (Park, 2019a, pp. 98–99).

## 4.3 The Argument from Undiscovered Evidence

Galileo observed the surface of the moon with the telescope that he invented; his observation disconfirmed the Ptolemaic theory. Such a historical episode motivates Wray (2016) to construct the argument from unconceived methods. It holds that unconceived methods disconfirmed some past theories, so they will also disconfirm some current theories. He states that "some of the changes of theory in the past have been due to changes in methodology, and some of the changes of theory in the future will also likely be due to changes in methodology" (Wray, 2016, pp. 372–373).

However, the concept of unconceived method can be used to support realism (Park, 2018a, pp. 57–61). Ever since special relativity was born in 1905, it has been tested again and again by new methods, and it has passed all of those tests. In 1971, for example, physicists ascertained the time dilation predicted by special relativity. Fast-flying jets traveled around the earth twice with atomic clocks on board. The clocks were compared with those that remained in their original place, which allowed the scientists to ascertain the time dilation. Such methods constitute the following list:

**The List of Unconceived Methods (Park, 2018a, p. 57)**

- Fast-flying jets and atomic clocks are used to ascertain the time dilation of a fast-moving object (Hafele & Keating, 1972).
- Whenever the global positioning system (GPS) works, it confirms the prediction of the special theory of relativity that time slows down as an object moves faster.
- Whenever a particle accelerator works, it confirms the prediction of the special theory of relativity that the mass of an object increases as it moves faster (Plettner et al., 2005).

Atomic clocks, fast-flying jets, global positioning systems, and particle accelerators did not exist before the twentieth century. They were later invented and then used to support special relativity. It probably follows that more such methods exist. They are currently unconceived, but they will be conceived later, and they will be used to justify special relativity. As a result, our philosophical offspring will have a stronger reason for believing that there are more such methods. Consequently, the number of such methods is infinitely large. In sum, "since the special theory of relativity has been repeatedly reinforced by new methods, it will continue to be reinforced by new methods indefinitely into the future" (Park, 2018a, p. 54). Therefore, "its probability will go above 99%, although it will not go above 100%" (Park, 2018a, p. 60), and "We are justified in believing that such a theory is true" (Park, 2018a, p. 59).

Objectors might say that current scientists cannot ideate the new methods that will support special relativity, so such methods do not exist. My first response to this possible objection is to contend that scientists cannot ideate the new methods, not because such methods do not exist but rather because scientists are similar to their predecessors who could not ideate old methods, such as global positioning systems and particle accelerators, that already supported special relativity (Park, 2018a, p. 58). This response makes use of Stanford's insight. He contends that scientists cannot entertain alternatives to current theories, not because alternatives do not exist but rather because they are similar to their predecessors "whose cognitive constitutions are not well suited to the task of exhausting the kinds of spaces of serious candidate theoretical explanations from which our scientific theories are drawn" (Stanford, 2006, p. 45).

My second response to the possible objection above is to invite readers to imagine that the objectors of the early twentieth century thought that there were no unconceived methods for supporting special relativity. We can see from the present perspective that they were wrong. Analogously, the objectors of the twenty-first century think that there are no unconceived methods for supporting special relativity. Our descendants of the twenty-second century will see from the future perspective that the objectors of the twenty-first century were wrong. This response utilizes Wray (2013) and Nickles's (2017) insight mentioned in Sect. 4.2 of this chapter.

In addition, atomic clocks, fast-flying jets, global positioning systems, and particle accelerators are significantly different from one another. It follows that hitherto unconceived methods will also be significantly different from one another and that each of them will significantly raise the probability that special relativity is true (Park, 2018a, p. 59). Moreover, they differ radically from currently conceived methods. This claim echoes Stanford's (2006, p. 19) and Wray's (2016, p. 369) insight that unconceived alternatives that will take the place of current theories differ radically from current theories, so current theories are not approximately true, contrary to what realists contend.

Similar arguments can be constructed for other scientific theories. For example, Charles Darwin (1859/1993) claimed that all present terrestrial organisms on the earth originated from one common ancestor. Molecular biologists discovered in the twentieth century that all current organisms use the same genetic code (Sober, 2018, p. 41). In addition, evolutionary theory claims that amphibians evolved from fish. Scientists hence looked for the fossils of intermediate organisms between fish and land animals with limbs. In 2004, the fossil of such an organism, Tiktaalik, was finally discovered in the Canadian Arctic (Shubin et al., 2006). Let me call such evidence "the undiscovered evidence." The genetic evidence and the fossil evidence were previously undiscovered, but were later discovered and then bolstered evolutionary theory. As we discover more such evidence, the inductive rationale will grow stronger for believing that there is more such evidence. It follows that the amount of such evidence is infinite and that the probability of evolutionary theory is destined to asymptotically approach 100%.

Note that I am not running circular arguments for special relativity and evolutionary theory. In other words, I am not claiming that special relativity and evolutionary theory are true, so they will continue to be reinforced by undiscovered evidence, and that they will continue to be reinforced by undiscovered evidence because they are true. I am rather running enumerative inductions for special relativity and evolutionary theory, i.e., I am saying that since special relativity and evolutionary theory were reinforced by previously undiscovered evidence, they will continue to be reinforced by hitherto undiscovered evidence.

These arguments for special relativity and evolutionary theory have the following form: since T has been supported by previously undiscovered evidence, it is likely[1] that it will continue to be supported by hitherto undiscovered evidence. Any argument of this form might be called "the argument from undiscovered evidence." The argument from undiscovered evidence echoes Stanford's problem of unconceived alternatives (PUA) that was examined in Subsection 2.4.2 of Chap. 2. According to the PUA, since scientists in the past lacked the cognitive capacity to come up with alternatives to their theories, current scientists also lack the cognitive capacity to come up with alternatives to their theories. In addition, Stanford asserts that the possibility space of rival theories "appears to be indeterminate and unbounded" (Stanford, 2006, p. 133). Just as Stanford adds this claim of infinity to the PUA, so advocates of the argument from undiscovered evidence add to their

[1] I drop "it is likely" hereafter for the sake of brevity, as I dropped it from the formulation of the classic PI in Sect. 2.2 of Chap. 2.

argument the claim that the amount of undiscovered evidence is infinite. The PUA and the argument from undiscovered evidence are in the same boat.

The argument from undiscovered evidence overcomes Florian Müller's (2015) and Mario Alai's (2017, p. 3283) objection to realism (Park, 2018a, pp. 59–60). In response to the classic PI, realists point out that present theories are better than the occupants of the graveyard of dead theories. Müller and Alai acknowledge realists' point. However, Müller contends that it "is not at all obvious why science, or at least our current best theories, should have achieved a degree of success that warrants their truth" (Müller, 2015, p. 406). Similarly, Alai says that it "is hard to think that any improvement of our background empirical knowledge and methods can at some point make scientists practically infallible; and even more that this point has already been reached" (Alai, 2017, p. 3283). Realists can happily admit that current theories are not successful enough to show that they are true, but they would argue that current theories are fated to be augmented by an infinite amount of undiscovered evidence, so they are warranted now.

How would objectors respond to the argument from undiscovered evidence? They might argue that it involves a hasty generalization, pointing out that there is an infinitely wide gap between the premise that T has been supported by previously undiscovered evidence and the conclusion that it will continue to be supported by hitherto undiscovered evidence.

My response to the foregoing objection is to contend that the argument from undiscovered evidence is no worse off than the PUA. The PUA appeals to Stanford's (2006, pp. 19–20) list cited in Subsection 2.4.6 of Chap. 2. The list indicates that Stanford takes the existence of the caloric theory and thermodynamic theories as evidence for the existence of infinitely many unconceived theories of heat. There is an infinitely wide gulf between the premise and the conclusion of the PUA. It follows that the PUA and the argument from undiscovered evidence are in the same boat (Park, 2018a, p. 60).

Pessimists might reply that the evidence for the existence of infinitely many theories of heat is not just the existence of the caloric and kinetic theories but rather the existence of all theories in science. For instance, the existence of the Ptolemaic theory and Aristotelian mechanics is also evidence for the existence of infinitely many theories of heat. In general, the evidence for the existence of unimagined rivals in a field of science comes not only from the existence of the imagined theories in that field of science but also from the existence of the imagined theories in other fields of science. Stanford's list is much longer than the list of undiscovered evidence for special relativity in Sect. 4.3 of this chapter. Accordingly, Stanford's PUA is much stronger than the argument from undiscovered evidence for special relativity.

However, advocates of the argument from undiscovered evidence would make a similar move. What makes it likely that there is an infinite amount of undiscovered evidence for special relativity is not only the existence of unconceived evidence, as mentioned in Sect. 4.3 of this chapter, but also the existence of all the undiscovered evidence in all of the fields of science. For example, DNA sequencing was undiscovered, was later discovered, and then supported evolutionary theory. Such evidence can be added to the list of undiscovered evidence for special relativity in Sect. 4.3 of this chapter. As a result, the list can be expanded ad nauseam by the previously undiscovered evidence in biology, chemistry, geology, psychology, and so on.

It is useful in this context to introduce two rival views in the realism debate called "localism" and "globalism." Localism holds that we should evaluate different fields of science separately and that we should not evaluate two or more fields of science together. By contrast, globalism holds that we should "evaluate all scientific theories or disciplines together" (Park, 2019c, p. 360). This definition of globalism goes along with Jamin Asay's definition, according to which globalism is "concerned with the well-confirmed theories that belong to any and all of the sciences" (Asay, 2019, p. 589).

Asay (2019) rejects globalism and accepts localism, mainly on the grounds that the different fields of science are epistemically distinct from each other. The evidence for the existence of mitochondria, for example, is different from the evidence for the existence of the Higgs field. Consequently, we should take different epistemic attitudes toward the theoretical entities of the different fields of science. Under the framework of localism, pessimists can construct "a successful but localized pessimistic induction" (Asay, 2019, p. 605). A local PI ranges only over a particular field of science, but not over any other fields of science. By contrast, a global PI ranges over all fields of science. All of the six first-order PIs that we discussed in Chap. 2, such as the classic PI and the PUA, are global PIs.

Although all of the six first-order PIs are developed under the globalist framework, they can be modified to fit the localist framework. For example, the globalist version of the classic PI states that since past theories in all fields of science were discarded, current theories in all fields of science will be discarded. The localist version of the classic PI states that since past theories in a particular field of science were abandoned, current theories in that field of science will be abandoned. All the other five first-order PIs can be similarly modified to conform to the localist framework. I do not spell out their localist versions in the interest of saving space.

The argument from undiscovered evidence, which was originally formulated under the globalist framework, can also be modified to conform to the localist framework. The localist version of the argument from undiscovered evidence says that since T in a particular field of science has been supported by previously undiscovered evidence, it will continue to be supported by hitherto undiscovered evidence. The localist version denies that the undiscovered evidence in other fields of science can also be evidence for the existence of hitherto undiscovered evidence that will support T.

Now, pessimists might make another attempt to undermine the argument from undiscovered evidence. Specifically, they might withdraw Stanford's contention that the possibility space of rivals "appears to be indeterminate and unbounded" (Stanford, 2006, p. 133). They might now contend that the number of unimagined alternatives to T is finite. Since they no longer claim that the number of unimagined alternatives is infinite, I can no longer claim that the amount of undiscovered evidence for T is infinite and thus that the argument from undiscovered evidence is no worse off than the PUA vis-à-vis the charge that they are hasty generalizations.

This move comes at a cost. As Samuel Ruhmkorff notes, "as long as there is a finite number of theories in a domain, the falsity of past theories will tend to increase rather than decrease our confidence in current theory" (Ruhmkorff, 2013, p. 412). Ruhmkorff's insight creates the following burden of proof for pessimists. If scientists had thrown out enough false alternatives, current theories would be true. If they had not thrown out enough false alternatives, current theories would be false. Scientists have either thrown out enough false alternatives or not. The PUA implies that current theories are false. Therefore, pessimists need to establish that "the probability that scientists have not yet eliminated enough false alternatives is higher than the probability that scientists have eliminated enough false alternatives" (Park, 2019d, p. 140).

This challenge to pessimists indicates that the more there are occupied places in the graveyard of dead theories, the better it is for realism and the worse it is for pessimism. Realists would be pleased to see the graveyard crowded with dead theories, thinking that there are already enough occupants in the graveyard, so current theories will not make it there. By contrast, pessimists would be pleased to see the graveyard uncrowded with dead theories, thinking that there are not enough occupants in the graveyard, so current theories will eventually make it there. Therefore, contrary to what pessimists assume, it is wrong for pessimists to strive for a long list of superseded theories (Park, 2019d, p. 141).

## 4.4 The Argument from Neighboring Theories

Michael Friedman (1981, p. 7), Peter Kosso (1992, p. 38), and Stathis Psillos (1999, p. 205) observe that two scientific theories that were confirmed independently of each other can jointly explain or predict new phenomena. Let me give the following example (Park, 2019b, p. 89). According to the theory of evolution, marsupials thrived in South America in the distant past. According to plate tectonics, Australia, Antarctica, and South America were different parts of the same continent in the distant past. Aware of these claims of evolutionary theory and plate tectonics, Michael Woodburne and William Zinsmeister (1982) hypothesized that marsupials moved from South America to Antarctica and then to Australia and hence that there are marsupial fossils in Antarctica. On the basis of this hypothesis, they excavated a region of Antarctica and discovered marsupial fossils. In this scientific episode, the two theories jointly retrodicted and explained new phenomena. One theory was unconceived, was later conceived, and then supported the other theory, and vice versa.

Evolutionary theory is similarly supported by other neighboring theories: molecular biology, the germ theory, and electromagnetic theory (Park, 2019b, pp. 89–90). In the interest of saving space, I do not spell out how evolutionary theory predicts or explains new phenomena in conjunction with each of these neighboring theories. I only cite the following list of the neighboring theories that have reinforced evolutionary theory:

**The List of Neighboring Theories**
- The theory of plate tectonics
- Molecular biology
- Electromagnetic theory
- The germ theory (Park, 2019b, p. 92)

This list entitles us to construct a philosophical argument for evolutionary theory, according to which "because evolutionary theory has been supported by its several neighboring theories that were previously unconceived, it will be supported by infinitely many hitherto unconceived neighboring theories" (Park, 2019b, p. 87).

We can construct similar arguments for other theories. For example, electromagnetic theory has been bolstered by evolutionary theory, molecular biology, and plate tectonics exactly in the way evolutionary theory has been bolstered by the germ theory, molecular biology, plate tectonics, and electromagnetic theory (Park, 2019b, pp. 90–91). In the interest of saving space, I do not specify how electromagnetic theory predicts or explains new phenomena in conjunction with these neighboring theories. I only note that the neighboring theories entitle us to construct an argument for electromagnetic theory, according to which electromagnetic theory was augmented by previously unimagined neighboring theories, so it will continue to be augmented by hitherto unimagined neighboring theories. The probability of electromagnetic theory being true will asymptotically approximate 100%. Hence, we are now justified in believing it.

The foregoing arguments for evolutionary theory and electromagnetic theory have the form: since T has been reinforced by previously unconceived neighboring theories, it will continue to be reinforced by hitherto unconceived neighboring theories. Any argument of this form might be called "the argument from neighboring theories" (Park, 2019b, p. 87). It echoes Stanford's PUA, just as the argument from undiscovered evidence echoes Stanford's PUA. The claim that the number of unconceived neighboring theories is infinite can be either added to or not added to the argument from neighboring theories, just as the claim that the amount of undiscovered evidence is infinite can be added to or not added to the argument from undiscovered evidence.

The argument from neighboring theories is distinct from the NMA. The former uses a rule of inference called "enumerative induction," whereas the latter uses inference to the best explanation (IBE) (Park, 2019b, p. 94). It is not liable to the objection (Laudan, 1981, p. 45; Fine, 1991, p. 82) that philosophers' use of IBE is just as problematic as scientists' use of IBE. In addition, the argument from neighboring theories mirrors Stanford's PUA, while the NMA does not. Therefore, future criticisms against the argument from neighboring theories will apply no less to the PUA.

In Sect. 4.4 of this chapter, we went over the objections and the replies between proponents and opponents of the argument from undiscovered evidence. There can be the same objections and replies mutatis mutandis between proponents and opponents of the argument from neighboring theories. For example, the opponents might say that Stanford's list is longer than the list of the neighboring theories that supported evolutionary theory. In response, the proponents could expand the list of neighboring theories by adding to it the neighboring theories that support other theories, such as electromagnetic theory and plate tectonics. I do not flesh out this debate between proponents and opponents of the argument from neighboring theories for the sake of saving space.

Realists can use the argument from neighboring theories to confront Müller's (2015, p. 406) and Alai's (2017, p. 3283) challenge to realism. Realists can happily grant that we do not know whether current theories are successful enough for us to believe that they are true. However, they can argue that current theories are destined to be bolstered by unconceived neighboring theories indefinitely into the future, so their probabilities will asymptotically approach 100%, and we are now justified in believing them.

How would pessimists object to the argument from neighboring theories? They might argue that the fact that two theories jointly predict or explain new events does not increase their probabilities at all, i.e., the joint prediction or explanation of new phenomena does not make the two theories more probable. Their probabilities are zero whether or not they jointly predict new events. Therefore, it is wrong to conclude that their probabilities will asymptotically approach 100% and that we can justifiably believe them now.

How do I respond to this possible objection from pessimists? I point out that pessimists have the burden of showing that enumerative induction is a reliable rule of inference, but that IBE is not (Park, 2019b, p. 96). The PI against scientific theories uses enumerative induction, which implies that pessimists take enumerative induction to be epistemically efficacious. The possible objection above implies that pessimists take IBE to be epistemically inefficacious. Therefore, pessimists have the burden of pointing out the relevant difference between enumerative induction and IBE that would entitle them to accept enumerative induction but reject IBE.

Antirealists might appeal to Humean skepticism and then say that both enumerative induction and IBE are unreliable rules of inference, so the argument from neighboring theories is not convincing to them. In response, I remind them that this move has many costs and only one benefit. The only benefit is that they can reject the argument from neighboring theories. The costs are that several criticisms that I leveled at antirealism in Chap. 3 apply mutatis mutandis to them. For example, epistemic reciprocalists will appeal to Humean skepticism to reject their positive theories, whatever they might be.

## 4.5 The Counter-Induction for Realism

We are inductionists in certain contexts but counter-inductionists in other contexts. For example, we observe that apples have fallen down and then infer that they will fall down in the future. By contrast, we observe that Tom has made mistakes, and then we infer that he will no longer make similar mistakes. Induction presupposes the uniformity principle that the future will be like the past (Hume, 1888/1978, p. 89). By contrast, counter-induction presupposes the disuniformity principle that the future will not be like the past (Park, 2017b, p. 213).

It is common knowledge that medical scientists go through many trials before they develop a new medicine to quell an epidemic. Imagine that they make a new material, $m_1$. They try it on white mice, and it does not work. They make another new material, $m_2$, and they try it on white mice, but it does not work either. Imagine that they make $m_3$, which has not yet been tested. Pessimists and counter-inductionists make opposing predictions about whether it will work. Pessimists argue that since $m_1$ and $m_2$ did not work, $m_3$ will not work either. This inference is based upon their fundamental belief that the future is similar to the past. By contrast, counter-inductionists argue that since $m_1$ and $m_2$ did not work, $m_3$ will work. This inference is based upon their fundamental belief that the future is not similar to the past. Medical scientists try $m_3$ on white mice, and it does not work. Pessimists point out that counter-inductionists' prediction about $m_3$ was false. Counter-inductionists reply that they did not predict that $m_3$ would definitely work, and that they rather predicted that it would probably work, so their prediction was compatible with the experimental outcome. Now, medical scientists make $m_4$, and they try it on white mice. Pessimists are more convinced that $m_4$ will not work, whereas counter-inductionists are more convinced that $m_4$ will work. It works, just as the counter-inductionists predicted. In this case, counter-induction works, and induction does not. A few concrete scientific episodes in which scientists had success after multiple failures can be found in Seungbae Park (2017b, p. 216, 2018b, p. 330).

The disagreement between pessimists and counter-inductionists can be extended to the realism debate. Pessimists contend that since earlier theories were false, their present successors are false too. Pessimists believe that past failures are an indication that there will be more failures and that the more theories were overturned, the more probable it is that current theories will be overturned. By contrast, counter-inductionists contend that "since past theories were false, present theories are true" (Park, 2018b, p. 330). As far as counter-inductionists are concerned, past failures are an indication of subsequent success, and the more theories were overturned, the more probable it is that current theories will not be overturned.

Suppose that present theories are ousted. Is the counter-induction then refuted? The answer is "No." Strictly speaking, the counter-induction does not say that current theories are *definitely* true. Rather, it says that they are *probably* true. Therefore, it is wrong for objectors to contend that the counter-induction will be refuted once current theories are overturned. As far as counter-inductionists are concerned, the demise of present theories will only make future theories more likely to be true. As a result, future counter-inductionists will be more convinced of the theories of their times than current counter-inductions are of current theories.

Let me elucidate an implication of the counter-induction. Pessimists are in a race toward producing longer and longer lists of obsolete theories. The lists of different lengths are provided by Larry Laudan (1981, p. 33) and Stanford (2006, pp. 19–20), as cited in Sect. 1.3 of Chap. 1 and Subsection 2.4.6 of Chap. 2, respectively. Pessimists believe that the longer the list is, the more probable it is that present theories are false. By contrast, counter-inductionists believe that the longer the list is, the more probable it is that present theories are true (Park, 2018b, p. 337).

The counter-induction for realism is immune to the following criticisms against the NMA. The success of science does not cry out for an explanation (Sect. 1.4, Chap. 1). The NMA begs the question against the critics of IBE (Sect. 1.5, Chap. 1). Realism is an ad hoc hypothesis (Sect. 1.8, Chap. 1). Science is successful because theories undergo natural selection (Sect. 1.6, Chap. 1), because successful theories are empirically adequate (Sect. 1.9, Chap. 1), because observables operate as if successful theories were true (Sect. 1.10, Chap. 1), and because successful theories are observationally similar to true theories (Sect. 1.11, Chap. 1). The NMA is vulnerable to these criticisms because it attempts to arrive at realism by explaining the success of science. By contrast, the counter-induction for realism uses an alternative route, viz., the history of science, to arrive at realism. Therefore, none of those criticisms applies to the counter-induction for realism.

The counter-induction for realism neutralizes Nickles's (2017) objection to realism. He contends that realism results from the end of history illusion, as we have seen in Sect. 1.7 of Chap. 1. According to realism, there will be few scientific revolutions in the future, although there were many scientific revolutions in the past. On Nickles's account, realists are those who "see the future of a currently successful field as similarly flat—as if the dynamical, innovative history of the field is now ended" (Nickles, 2017, p. 152). Counter-inductionists would say that since there have been many theoretical changes in the past, there will be few theoretical changes in the future and that it is wrong to think that since there were many theoretical changes in the past, there will be many theoretical changes in the future.

Laudan and Timothy Lyons object that outdated theories are counterexamples to realism, as we saw in Sect. 1.3 of Chap. 1. Counter-inductionists would reply that such theories are not counterexamples to realism but are, in fact, evidence for realism. Counter-inductionists are thankful to past scientists for repeatedly replacing their theories with alternatives. Without their fortitude to obtain true theories despite repeated failures, counter-inductionists would have only a flimsy basis for the counter-induction for realism. On counter-inductionists' account, what antirealists need to show to rebut realism is not that there were many theoretical upheavals, but that there were few theoretical upheavals in the history of science.

Counter-inductionists are appalled when called upon to imagine what would have happened in the history of science if scientists had been pessimists. Scientists would have been less motivated to develop alternatives to their accepted theories, thinking that alternatives would be thrown into the graveyard of discarded theories, just as the predecessors of their accepted theories had been. The PI demotivates scientists to be creative and develop alternatives (Park, 2016, p. 121). Therefore, pessimism is a useless philosophical framework for scientists.

Counter-inductionists are thrilled to imagine what would have happened in the history of science if scientists had been counter-inductionists. Scientists would have been motivated to develop alternatives to the theories of their time, thinking that since the theories of their time had run into anomalies, alternatives to them would not and that since the theories of their time would be thrown into the graveyard of discarded theories, alternatives to them would not be. Unlike the PI, the counter-induction motivates scientists to be creative and develop alternatives to existing theories. Therefore, it is a useful philosophical framework for scientists.

One caveat is in order. I do not claim that we should be counter-inductionists in all contexts. Rather, I claim that we should be inductionists in some contexts and counter-inductionists in others. Therefore, it is a mistake to attempt to refute my position by presenting a few examples in which induction works (Park, 2018b, p. 340).

In Sect. 4.1 of this chapter, I claimed that all of the six new arguments for realism developed in this chapter utilize the same theoretical machinery that sharp antirealists have developed to attack realism, which implies that even the counter-induction for realism utilizes antirealists' idea. What exactly is antirealists' idea? It is that we should look into the history of science in order to foresee the fate of current theories. This idea is built into both the PI and the counter-induction.

This idea would be rejected by Peter Lipton, who says that at "the end of the day, the only evidence for the truth of scientific theories is the evidence that scientists' use, and the only positive arguments for scientific realism are the arguments that scientists make" (Lipton, 2001, p. 353). On Lipton's account, realists should turn to scientists' arguments instead of the NMA for a defense of realism. This insight is endorsed by Peter Achinstein (2002), Patrick Enfield (2008), Sherrilyn Roush (2010), Simon Fitzpatrick (2013), and Park (2017a, p. 99, 2019c, pp. 362–364). These philosophers would say that antirealists should expose intrinsic flaws with scientists' arguments for current theories instead of looking into the history of science, if they aim to take the antirealist attitude toward current theories.

## 4.6 The English Notion of Rationality

Bas van Fraassen (1989) develops two competing notions of rationality. One is the Prussian notion, which holds that "what is rational to believe is exactly what one is rationally compelled to believe" (van Fraassen, 1989, p. 171). The other is that the English notion, which holds that "what it is rational to believe includes anything that one is not rationally compelled to disbelieve" (van Fraassen, 1989, pp. 171–172). The Prussian notion parallels the Prussian legal system, according to which "everything is forbidden which is not explicitly permitted" (van Fraassen, 1989, p. 171). The English notion parallels the English legal system, according to which "everything is permitted that is not explicitly forbidden" (van Fraassen, 1989, p. 171).

Van Fraassen adopts the English notion, whose alternative formulation is that "what it is rational to disbelieve includes anything that one is not rationally compelled to believe" (Park, 2020, p. 2). Van Fraassen uses this formulation of the

English notion to defend his disbelief of T. He argues that IBE is not rationally[2] compelling, so even if T best explains certain events, it is not compelling, and thus under the English notion, it is rational to disbelieve T. He also argues that we have no compelling argument for disbelieving T, so under the English notion, it is rational to believe T. In sum, it is rational for realists to believe T, but it is also rational for antirealists to disbelieve T. Keep in mind that under the English notion, both realists' belief and antirealists' disbelief of T are rational.

Van Fraassen disbelieves T, saying that the belief of T "is supererogatory" (van Fraassen, 2007, p. 343, 2017, p. 99). To say so means that it is beyond the call of rationality to believe T. In other words, it is rational to believe T, but we do not need to believe T to be rational. Thus, it is rational to believe T, but also rational not to believe T. This cryptic claim can be illuminated by the following analogy. It is supererogatory for you to donate your kidney to a hospital. To say so means that morality does not require that you do so. It is moral for you to donate your kidney, but you do not need to do it to be moral. Thus, it is moral for you do it, but it is also moral for you not to do it. Just as it is supererogatory to donate a kidney, so it is supererogatory to believe T.

Realists cannot refute van Fraassen's position by establishing that it is rational to believe T, for he readily admits that it is. Suppose that the NMA is cogent, so it is rational to believe T. It does not follow that van Fraassen's position is refuted. The refutation of his position does not require the thesis that it is rational to believe T, but rather requires the thesis that it is irrational to disbelieve T, i.e., "it is unreasonable not to believe that H is true" (van Fraassen, 2017, p. 102). To use the analogy above, the refutation of the position that it is supererogatory for you to donate your kidney does not require the thesis that it is moral for you to donate your kidney, but rather requires the thesis that it is immoral for you not to donate your kidney.

How should we assess van Fraassen's foregoing case for antirealism? In my view, realists can make a similar move, i.e., they can also invoke the English notion to defend the belief of T (Park, 2019e, pp. 479–480). They believe T, while arguing that we have no compelling argument to believe or to disbelieve T, so both realists' belief of T and antirealists' disbelief of T are reasonable. The refutation of realists' position does not require the thesis that it is rational to disbelieve T, but rather requires the thesis that it is irrational to believe T.

It is supererogatory for antirealists to criticize realists' belief of T, for they believe that realists' belief of T is reasonable. It is also supererogatory for realists to criticize antirealists' disbelief of T, for they believe that antirealists' disbelief of T is reasonable. Antirealists can rationally settle for the disbelief of T without criticizing realists' belief of T. Analogously, realists can rationally settle for the belief of T without criticizing antirealists' disbelief of T. Armed with the English notion of rationality, realists and antirealists are on even footing.

The foregoing argument for realism might be called "the English argument for realism." It is so named because it invokes the English notion of rationality to

[2] I drop "rationally" hereafter for the sake of brevity.

establish that it is rational to believe T. It is analogous to van Fraassen's English argument for antirealism. His argument invokes the same notion of rationality to establish that it is rational to disbelieve T. His argument might be called "the English argument for antirealism." The English arguments for realism and antirealism rely on the same notion of rationality, and thus they are in the same boat.

The English argument for realism is useful in overcoming Müller's (2015) and Alai's (2017) objection to realism that realists have the burden of showing that unlike past theories, current theories are successful enough to show that they are true. The degree of success of past theories was below a certain point that I called "the flipping point" in Subsection 2.4.3 of Chap. 2. The flipping point is the point at which unobserved instances become different from observed instances. Realists, invoking the English notion of rationality, can reply that we have no compelling argument to disbelieve that the degree of success is above the flipping point, so it is rational to believe that the degree of success is above the flipping point. However, they welcome antirealists to disbelieve that the degree of success is above the flipping point. They admit that we have no compelling argument to believe that the degree of success is above the flipping point and that in the absence of such an argument, it is rational for antirealists to disbelieve that the degree of success is above the flipping point (Park, 2020, pp. 6–7).

The English argument for realism is also useful in overcoming van Fraassen's (1989, p. 143) argument from a bad lot, which we discussed in Sect. 1.5 of Chap. 1. It says that realists need to show that the true hypothesis is more likely to exist in the collection of considered hypotheses than in the collection of unconsidered hypotheses before they believe that the best of the considered hypotheses is true. Realists believe, whereas antirealists disbelieve, that the true hypothesis is more likely to exist in the collection of imagined hypotheses. The English notion can help realists in this standoff. They can argue that we have no compelling argument to disbelieve that the true hypothesis is more likely to exist in the collection of imagined hypotheses and that in absence of such an argument, it is rational for realists to believe that the true hypothesis is more likely to exist in the collection of imagined hypotheses. However, they welcome antirealists to disbelieve that the true hypothesis is more likely to exist in the collection of considered alternatives. They admit that we have no compelling argument to believe that the true hypothesis is more likely to exist in the collection of considered alternatives and that in the absence of such an argument, it is rational for antirealists to disbelieve that the true hypothesis is likely to exist in the collection of considered alternatives.

Stanford would object that he has a compelling argument to disbelieve that the true hypothesis is more likely to exist in the collection of imagined hypotheses. His argument is that "the history of scientific inquiry itself offers a straightforward rationale for thinking that there typically are alternatives to our best theories equally well-confirmed by the evidence" (Stanford, 2006, p. 20). Therefore, we are compelled to disbelieve that the true hypothesis is more likely to exist in the collection of imagined hypotheses, and it is irrational for realists to believe that the true hypothesis is more likely to exist in the collection of imagined hypotheses.

On close examination, however, the English notion can help realists confront the PUA. Suppose that there are a finite number of alternatives. Realists believe that theorists have ousted enough false competitors to conclude that they have reached true theories. By contrast, antirealists disbelieve that theorists have ousted enough false rivals to conclude that they have reached true theories. In other words, realists believe that Stanford's list cited in Subsection 2.4.6 of Chap. 2 is long enough to conclude that current theories are true, whereas antirealists disbelieve that it is long enough to conclude that current theories are true (Park, 2019d, p. 140). In this standoff, realists can invoke the English notion of rationality to argue that we do not have a compelling argument to disbelieve that theorists have thrown out enough false competitors and that in the absence of such an argument, it is rational for realists to believe that theorists have thrown out enough false rivals. However, they welcome antirealists to disbelieve that theorists have ousted enough false rivals, pointing out that we have no compelling argument to believe that theorists have ousted enough false rivals. They admit that in the absence of such an argument, it is rational for antirealists to disbelieve that theorists have ousted enough false rivals (Park, 2020, p. 7).

Suppose instead that there are an infinite number of competitive theories. Under this condition, the possibility space never shrinks no matter how many of a finite number of alternatives scientists may remove from it. It appears, therefore, that scientists will never obtain a true theory and that they are only fated to remove false theories in vain from the possibility space of alternatives.

A flaw with this reasoning becomes obvious once we consider the fact that scientists sometimes make achievements after engaging in multiple failures, as we saw in Sect. 4.5 of this chapter. There is the possibility space of ineffective materials that would not aid in an epidemic, and there is also the possibility space of effective materials that would aid in an epidemic. Scientists selected $m_1$, $m_2$, and $m_3$ from the former space and then selected $m_4$ from the latter space. Whenever scientists make an achievement after trials and errors, we can say that they moved from the possibility space of failures to the possibility space of achievements (Park, 2019d, pp. 141–142). It is wrong to think that once scientists eliminate some ineffective materials from the possibility space of ineffective materials, they are stuck in that possibility space, and they are only fated to eliminate ineffective materials indefinitely.

A similar criticism can be raised against the suggestion that scientists should rule out infinitely many *false* theories before they find a true theory. Scientists only need to *move* from the possibility space of infinitely many false theories to that of true theories in order to obtain the true theory. It is wrong to think that once scientists rule out some false alternatives from the possibility space of infinitely many false alternatives, they are stuck in that possibility space, and they are only destined to rule out false theories indefinitely.

There can be a dispute between realists and pessimists over whether scientists moved from the possibility space of infinitely many false theories to that of true theories in the history of science. Pessimists would say, for example, that oxygen theorists selected the oxygen theory from the space of infinitely many false theories, just as phlogiston theorists did. By contrast, realists would say that oxygen theorists switched spaces from the possibility space of infinitely many false theories to that of true theories and selected the oxygen theory from the latter space. Hence, realists and pessimists are in a stalemate (Park, 2019d, p. 141).

In this stalemate, realists can invoke the English notion of rationality, arguing that no argument compels us to disbelieve that oxygen theorists switched possibility spaces and that in the absence of such an argument, it is rational for realists to believe that oxygen theorists switched possibility spaces. However, they welcome pessimists to disbelieve that oxygen theorists switched possibility spaces. Realists admit that no argument compels us to believe that oxygen theorists switched possibility spaces and that in the absence of such an argument, it is rational for pessimists to disbelieve that oxygen theorists switched possibility spaces (Park, 2020, p. 8).

As we noted above, van Fraassen claims, appealing to the English notion of rationality, that it is rational for realists to believe T and also rational for antirealists to disbelieve T. This claim should be surprising to antirealists. Antirealists have raised many objections to realism, as we have seen in Chaps. 1 and 2. Van Fraassen's claim implies that none of those objections to realism compels us to disbelieve T and that none of them has enough force to make the belief of T unreasonable. To state this differently, we can rationally believe T despite the standard objections to realism, including the PI, which is alleged to be the most powerful argument against realism. It is my observation that van Fraassen has unwittingly provided realists with a magical way to get around the traditional objections to realism (Park, 2020, pp. 3–5).

## 4.7 The Grand Optimistic Induction

In Sect. 4.6, I provided five new arguments for realism that are distinct from the NMA. They can serve as the basis for a second-order argument for realism, which I call "the grand optimistic induction." It holds that since realism has been bolstered by previously unconceived arguments, it will continue to be bolstered by hitherto unconceived arguments. The premise of this argument is backed up by the following list:

**The List of New Arguments for Realism**
- The optimistic induction over realists
- The argument from undiscovered evidence
- The argument from neighboring theories
- The counter-induction for realism
- The English argument for realism

These five arguments were unconceived in the twentieth century, but they were conceived in the twenty-first century. They entitle us to infer that there are more such arguments, although they are presently unconceived. They will later be conceived and added to this list. As a result, our philosophical offspring will have a longer list of new arguments for realism, and they will have a stronger reason for inferring that there are unconceived arguments for realism. Thus, the list is fated to be expanded indefinitely into the future.

Pessimists might object that present realists cannot conceive of the new arguments for realism, so such arguments do not exist. My rejoinder to this possible objection is that present realists cannot conceive of the new arguments not because they do not exist, but rather because present realists' cognitive capacity is similar to that of the realists of the twentieth century who could not conceive of the five new arguments on the list.

The grand optimistic induction predicts that future realists will construct hitherto unconceived new arguments for realism. I ask readers to keep an eye on whether this prediction will follow the fate of the prediction of the grand PI mentioned in Subsection 2.4.6 of Chap. 2. Recall that the grand PI predicted that pessimists would construct a new PI, a successor to the PUA, and that as the grand PI predicted, Greg Frost-Arnold (2019) constructs the PME, a successor to the PUA.

## 4.8 Tu Quoque

The six new arguments for realism accord with epistemic reciprocalism, epistemic imperative, and the Golden Rule. Van Fraassen uses the English notion of rationality to defend his disbelief of T. In response, realists use the English notion of rationality to defend their beliefs of T. Realists welcome others, including van Fraassen, to use the English notion of rationality. In addition, pessimists act on the maxim, "Look at how theories have behaved to predict how they will behave." In accordance with this very maxim, realists construct the optimistic induction over realists, the argument from undiscovered evidence, the argument from neighboring theories, and the counter-induction for realism. Realists cheerfully follow pessimists' lead, acting on the maxim that pessimists act on.

The six new arguments for realism utilize antirealists' ideas, and thus they might prove to be the useful means for breaking the impasse between realists and antirealists. As Moti Mizrahi (2020, p. 13) observes, the realism debate is intractable because realists start from premises that antirealists find disagreeable and vice versa. For example, the NMA uses IBE to establish realism, but antirealists take IBE to be dubious. My response to the stalemate is to say that unlike the NMA, the six new arguments rely on antirealists' presuppositions, e.g., enumerative induction is reliable. Hence, antirealists should find the six new arguments for realism convincing.

Objectors might argue that I have committed *tu quoque*, pointing out that we cannot establish that an argument is correct by establishing that it is similar to an incorrect argument. On this account, antirealists' arguments against realism are all weak, and thus to say that my six new arguments for realism utilize antirealists' ideas only proves that they are similarly weak.

How do I respond to this *tu quoque* objection? Antirealists cannot raise it because it implies that their best arguments for antirealism are incorrect. What about those on the fence, those who are observing the debate between realists and antirealists? If they endorse the *tu quoque* objection, they would have to accept realism and reject antirealism. As noted in Sect. 1.1 of Chap. 1, the NMA and the PI are the main considerations for realism and antirealism, respectively. The *tu quoque* objection implies that the most forceful argument for antirealism, viz., the PI, is removed from the table. By contrast, the most forceful argument for realism, viz., the NMA, remains on the table. Moreover, antirealism has several epistemic and pragmatic disadvantages, as noted in Chap. 3. For these reasons, those on the fence, if they endorse the *tu quoque* objection, would have to lean toward realism.

Let me make one more critical comment on the *tu quoque* objection. Instead of dismissing the six new arguments and their antirealist counterparts as weak, we might as well explore how they relate to one another. For example, how does van Fraassen's English argument for antirealism relate to the argument from unconceived evidence and the argument from neighboring theories? As far as I can tell, it conflicts with them. Van Fraassen's position is supported by the English argument for antirealism, and thus the refutation of it requires the thesis that "it is unreasonable not to believe that H is true" (van Fraassen, 2017, p. 102). The thesis can be established by the argument from unconceived evidence and the argument from neighboring theories. These two arguments assert that once T has been supported by previously undiscovered evidence or by previously unconceived neighboring theories, its probability is destined to asymptotically approach 100%. It is unreasonable not to believe theories whose probabilities are destined to asymptotically approach 100%.

## 4.9 Summary

It is an open question whether the six new arguments for realism are tenable. However, they are no worse off than their antirealist counterparts. Specifically, the optimistic induction over realists is no better off than the PI over realists. The argument from undiscovered evidence and the argument from neighboring theories are no worse off than the PUA. The counter-induction for realism is no worse off than the classic PI. The English argument for realism is no worse off than van Fraassen's English argument for antirealism. The grand optimistic induction is no worse off than the PUA. I predict that antirealists' future objections to these six new arguments for realism will apply no less to their antirealist counterparts. In sum, antirealists have provided realists with the magical means of defending realism. To reject this conclusion is to reject antirealists' arguments for antirealism.

## References

Achinstein, P. (2002). Is there a valid experimental argument for scientific realism? *The Journal of Philosophy, 99*(9), 470–495.

Alai, M. (2017). Resisting the historical objections to realism: Is Doppelt's a viable solution? *Synthese, 194*(9), 3267–3290.

Asay, J. (2019). Going local: A defense of methodological localism about scientific realism. *Synthese, 196*(2), 587–609.

Chakravartty, A. (2008). What you don't know can't hurt you: Realism and the unconceived. *Philosophical Studies, 137*(1), 149–158.

Darwin, C. (1859/1993). *The portable Darwin* (P. Graham, Ed.). Penguin Books.

Enfield, P. (2008). Review. Exceeding our grasp: Science, history, and the problem of unconceived alternatives. K. Stanford. *The British Journal for the Philosophy of Science, 59*(4), 881–895.

Fine, A. (1991). Piecemeal realism. *Philosophical Studies, 61*(1–2), 79–96.

Fitzpatrick, S. (2013). Doing away with the no miracles argument. In V. Karakostas & D. Dieks (Eds.), *EPSA11 perspectives and foundational problems in philosophy of science* (Vol. 2, pp. 141–151). The European Philosophy of Science Association Proceedings, Springer.

Friedman, M. (1981). Theoretical explanation. In R. Healey (Ed.), *Reduction, time & reality: Studies in the philosophy of natural science* (pp. 1–16). Cambridge University Press.

Frost-Arnold, G. (2019). How to be a historically motivated antirealist: The problem of misleading evidence. *Philosophy of Science, 86*(5), 906–917.

Hafele, J., & Keating, R. (1972). Around-the-world atomic clocks: Predicted relativistic time gains. *Science, 177*(4044), 166–168.

Hume, D. (1888/1978). *A treatise of human nature* (L. A. Selby-Bigge & P. H. Nidditch, Eds.). Oxford University Press.

Kosso, P. (1992). *Reading the book of nature: An introduction to the philosophy of science*. Cambridge University Press.

Laudan, L. (1981). A confutation of convergent realism. *Philosophy of Science, 48*(1), 19–49.

Lipton, P. (2001). Quests of a realist. *Metascience, 10*(3), 347–353.

Mizrahi, M. (2020). *The relativity of theory: Key positions and arguments in the contemporary scientific realism/antirealism debate*. Springer.

Müller, F. (2015). The pessimistic meta-induction: Obsolete through scientific progress? *International Studies in the Philosophy of Science, 29*(4), 393–412.

Nickles, T. (2017). Cognitive illusions and nonrealism: Objections and replies. In E. Agazzi (Ed.), *Varieties of scientific realism: Objectivity and truth in science* (pp. 151–163). Springer.

Park, S. (2011). A confutation of the pessimistic induction. *Journal for General Philosophy of Science, 42*(1), 75–84.

Park, S. (2016). How to foster scientists' creativity. *Creativity Studies, 9*(2), 117–126.

Park, S. (2017a). Selective realism *versus* individual realism for scientific creativity. *Creativity Studies, 10*(1), 97–107.

Park, S. (2017b). The uniformity principle vs. the disuniformity principle. *Acta Analytica, 32*(2), 213–222.

Park, S. (2018a). Justifying the special theory of relativity with unconceived methods. *Axiomathes, 28*(1), 53–62.

Park, S. (2018b). The anti-induction for scientific realism. *Grazer Philosophische Studien, 95*(3), 329–342.

Park, S. (2019a). Optimistic realism over selectivism. *Kriterion: Journal of Philosophy, 31*(1), 89–106.

Park, S. (2019b). The coherence of evolutionary theory with its neighboring theories. *Acta Biotheoretica, 67*(2), 87–102.

Park, S. (2019c). Localism vs. individualism for the scientific realism debate. *Philosophical Papers, 48*(3), 359–377.

Park, S. (2019d). New objections to the problem of unconceived alternatives. *Filosofia Unisinos, 20*(2), 138–145.

Park, S. (2019e). How to formulate scientific realism and antirealism. *Journal for General Philosophy of Science, 50*(4), 477–488.

Park, S. (2020). How to overcome antirealist objections to scientific realism. *Axiomathes, 30*(1), 1–12.

Plettner, T., Byer, R., & Siemann, R. (2005). The impact of Einstein's theory of special relativity on particle accelerators. *Journal of Physics B: Atomic, Molecular and Optical Physics, 38*, S741–S752.

Psillos, S. (1999). *Scientific realism: How science tracks truth.* Routledge.

Psillos, S. (2018). Tolstoy's argument: Realism and the history of science. *Spontaneous Generations: A Journal for the History and Philosophy of Science, 9*(1), 68–76.

Putnam, H. (1975). *Mathematics, matter and method: Philosophical papers, volume 1*. Cambridge University Press.

Roush, S. (2010). Optimism about the pessimistic induction. In P. Magnus & J. Busch (Eds.), *New waves in philosophy of science* (pp. 29–58). Palgrave Macmillan.

Ruhmkorff, S. (2013). Global and local pessimistic meta-inductions. *International Studies in the Philosophy of Science, 27*(4), 409–428.

Sober, E. (2018). *Philosophy of biology*. Westview Press.

Stanford, K. (2006). *Exceeding our grasp: Science, history, and the problem of unconceived alternatives*. Oxford University Press.

Stanford, K. (2009). Grasping at realist straws. *Metascience, 18*(3), 379–390.

Stanford, K. (2018). A fond farewell to 'approximate truth'? *Spontaneous Generations: A Journal for the History and Philosophy of Science, 9*(1), 78–81.

Shubin, N., Daeschler, E., & Jenkins, F. (2006). The pectoral fin of *Tiktaalik roseae* and the origin of the tetrapod limb. *Nature, 440*(7085), 764–771.

van Fraassen, B. (1989). *Laws and symmetry*. Oxford University Press.

van Fraassen, B. (2007). Reply: From a view of science to a new empiricism. In B. Monton (Ed.), *Images of empiricism: Essays on science and stances, with a reply from bas van Fraassen* (pp. 337–383). Oxford University Press.

van Fraassen, B. (2017). Misdirection and misconception in the scientific realism debates. In E. Agazzi (Ed.), *Varieties of scientific realism: Objectivity and truth in science* (pp. 95–108). Springer.

Vickers, P. (2017). Understanding the selective realist defence against the PMI. *Synthese, 194*(9), 3221–3232.

Vickers, P. (2018). Quo Vadis selective scientific realism. *Spontaneous Generations: A Journal for the History and Philosophy of Science, 9*(1), 118–121.

Woodburne, M., & Zinsmeister, W. (1982). Fossil land mammal from Antarctica. *Science, 218*(4569), 284–286.

Worrall, J. (1989). Structural realism: The best of both worlds. *Dialectica, 43*(1–2), 99–124.

Wray, B. (2013). The pessimistic induction and the exponential growth of science reassessed. *Synthese, 190*(18), 4321–4330.

Wray, B. (2016). Method and continuity in science. *Journal for General Philosophy of Science, 47*(2), 363–375.

# Chapter 5
# Critiques of Five Variants of Putative Realism

**Abstract** Entity realism, minimal realism, axiological realism, van Fraassen's definition of scientific realism, and selective realism all aspire to be versions of realism, as their names suggest. On close examination, however, they are not variants of realism but rather variants of putative realism. In addition, they are so close to antirealism that they are vulnerable to the objections that I raised against antirealism in Chap. 3, and as such their adherents have the burden of refuting the six new arguments for realism outlined in Chap. 4. The argument from a bad lot cannot be an argument for rejecting realism and accepting the five variants of putative realism.

**Keywords** Attributive realism · Axiological realism · Bad lot · Entity realism · Minimal realism · Putative realism · Selective realism

## 5.1 Introduction

Arthur Fine states that "there are altogether too many realisms" (Fine, 1986, p. 149). In a similar vein, Anjan Chakravartty says that "scientific realism is characterized differently by every author who discusses it, and this presents a challenge to anyone hoping to learn what it is" (Chakravartty, 2017). As these eminent philosophers suggest, many philosophical positions in the literature go by the name "realism." My reaction to those positions is to say that only one of them deserves the label. It says that "the theories accepted in a mature science are typically approximately true" (Putnam, 1975, p. 73). All the rest are variants of putative realism, i.e., they are generally accepted to be variants of realism.

What is the philosophical significance of the variants of putative realism? Some putative realists might argue that their positions are in the middle between realism and antirealism, and thus they can cheerfully reject both the six new arguments for realism and their antirealist counterparts that we discussed in Chap. 4. They can also accuse me of having committed *tu quoque*, i.e., they can say that it is wrong for me to attempt to establish realism by constructing arguments similar to antirealists'


S. Park, *Embracing Scientific Realism*, Synthese Library 445,
https://doi.org/10.1007/978-3-030-87813-9_5

weak arguments. In this chapter, I investigate whether the variants of putative realism are well motivated, and whether they are better than realism and antirealism.

Due to a limited amount of space, I am prevented from discussing all of the variants of putative realism found in the literature. As a result, I focus on five of them. They are entity realism (Hacking, 1983), minimal realism (Saatsi, 2019), axiological realism (Lyons, 2005, p. 167, 2016a, p. 564, 2018, p. 148, 2019, p. 833), scientific realism (van Fraassen, 1980), and selective realism (Psillos, 1999; Vickers, 2017). I already feel the resistance from many readers who protest that these views are forms of realism rather than forms of putative realism. These readers can rest assured that their resistance will subside once they finish reading this chapter. In each section below, I lay bare how little each of the five variants of putative realism is committed to the claims of science, i.e., how close each of them is to antirealism and, as a result, how disadvantageous each of them is. I also argue that the five variants of putative realism are not better off than realism vis-à-vis the argument from a bad lot.

## 5.2 Entity Realism

Entity realism is an original position that Ian Hacking (1983) stakes out. It holds that if we can manipulate certain theoretical entities, we can believe that they are real. He advances the famous slogan: "if you can spray them, then they are real" (Hacking, 1983, p. 23). This slogan applies to theoretical entities, such as electrons and protons, given that they can be manipulated with the use of scientific instruments like electron guns and particle accelerators.

Entity realism contrasts with theory realism, which holds that we can believe that a theory is true. Hacking rejects theory realism, and thus he does not believe that theoretical entities have the properties that a theory attributes to them. Let me derive an implication from rejecting theory realism. Quantum mechanics tells us that an electron acts like a wave when we do not measure it, but that it acts like a particle when we measure it. The problem of answering the question of what is involved in measurement is known as "the measurement problem." The measurement problem is a problem for theory realists, but not for entity realists, because theory realists believe, whereas entity realists do not, that electrons have the aforementioned properties.

Entity realism is silent not only about what properties theoretical entities have but also about how observational entities behave. Hence, it is silent about whether the predictions of a theory are true. It is one thing to say that theoretical entities exist. It is entirely another to say that a theory postulating theoretical entities makes true predictions. Thus, entity realism is less committed to what a theory says about observables than instrumentalism. This position might as well be called "entitism" rather than "entity realism" (Park, 2020a, p. 64). Alan Musgrave would concur with me on this count, given that he makes the following apposite comment concerning entity realism: "To believe in an entity, while believing nothing else about that entity, is to believe nothing or next to nothing" (Musgrave, 2017, p. 88).

What is wrong with believing next to nothing? In general, the less you believe, the less you can explain. For this reason, I put forward the slogan: "Believe less. Explain less" (Park, 2017a, p. 113). This point applies to entity realism. Entity realists believe that oxygen is real, but not that the oxygen theory is true. Consequently, they cannot use the oxygen theory to explain and predict events. If they do, they run into one of the three problems: Moore's paradox, the problem of disconcerting questions, and the problem of deceptive speech acts. Moreover, entity realism cannot provide an account of why a theory is successful. The mere hypothesis that oxygen is real cannot explain why the oxygen theory is successful. Musgrave claims that since entity realism lacks explanatory power, it is "a hopeless form of realism" (Musgrave, 2017, p. 88). In my view, entity realism is not a version of realism at all, but rather a version of putative realism. In any event, believing next to nothing comes with the disadvantage of explaining and predicting next to nothing.

Hacking's rejection of theory realism is predicated on Bas van Fraassen's contextual theory, according to which an "explanation is an answer to a why-question" (van Fraassen, 1980, p. 134) and "among the scientifically relevant factors, context determines explanatorily relevant ones" (van Fraassen, 1980, p. 126). Hacking states his appeal to the contextual theory as follows:

> Explanations are relative to human interests. I do not deny that explaining—'feeling the key turn in the lock' as Peirce put it—does happen in our intellectual life. But that is largely a feature of the historical or psychological circumstances of a moment. … Feeling the key turn in the lock makes you feel that you have an exciting new idea to work with. It is not a ground for the truth of the idea. (Hacking, 1983, p. 53)

How should we evaluate Hacking's appeal to van Fraassen's contextual theory? A dilemma can be constructed against Hacking (Park, 2017b, p. 62). The contextual theory is either warranted or unwarranted. If it is warranted, then scientific theories are also warranted, given that they are the products of inference to the best explanation (IBE). Since scientific theories are warranted, it is wrong for Hacking to reject theory realism. However, if the contextual theory is unwarranted, it is wrong for Hacking to appeal to it to reject theory realism, i.e., his rejection of theory realism is unwarranted. On either horn of this dilemma, it is wrong for him to reject theory realism.

Hacking might argue against the first horn of this dilemma in the following way. Philosophy and science are different intellectual enterprises, so they should be judged by different standards. Specifically, IBE is reliable when an explanans is a philosophical hypothesis, but it is not reliable when an explanans is a scientific hypothesis. Therefore, the contextual theory is warranted, whereas scientific theories are not.

However, scientists would reply that IBE in which a scientific theory is an explanans is reliable, whereas IBE in which a philosophical theory is an explanans is not. Therefore, a scientific theory is warranted, but the contextual theory is not (Park, 2017b, p. 62). If philosophers of science maintain that scientists should not engage in a double standard against their philosophical theories, they should not engage in a double standard against scientific theories. Again, there "is no reason for thinking that the Golden Rule ranges over moral matters, but not over epistemic matters" (Park, 2018a, pp. 77–78).

Hacking might now decide to abandon the contextual theory and adopt the pessimistic induction (PI) as his new route to the rejection of theory realism. However, this new route has many roadblocks. Imagine that caloric theorists thought that caloric was real because they manipulated caloric. We can now see that they were wrong. Kinetic theorists think that molecules are real because they manipulate them. As a matter of induction, our philosophical heirs will think that kinetic theorists were wrong. Thus, entity realism is no better off than theory realism vis-à-vis the PI.

Entity realists might retort that the kinetic theory is better confirmed than the caloric theory, and accordingly caloric theorists were wrong to think that they manipulated caloric, but kinetic theorists are right to think that they manipulate molecules. However, this retort presupposes that theory realism offers a means of arriving at entity realism. It goes against Hacking's original spirit that we can believe in theoretical entities without having the means of describing theoretical entities.

Furthermore, entity realists' foregoing attempt would fall prey to Florian Müller's (2015) and Mario Alai's (2017) objection discussed in Sect. 4.3 of Chap. 4. Recall that it "is not at all obvious why science, or at least our current best theories, should have achieved a degree of success that warrants their truth" (Müller, 2015, p. 406). The fact that the kinetic theory is better confirmed than the caloric theory gives us a reason for rejecting the pessimistic inference that since it transpired that caloric did not exist, it will transpire that molecules do not exist. But it does not amount to a positive argument for the belief that molecules are real.

A reviewer objects that I fail to come to grips with Hacking's discussion of the causal knowledge that experimental scientists arrive at about the experimental entities that they manipulate in their laboratories. Experimental scientists arrive at knowledge of the causal behavior of experimental entities, which entitles them to make true claims about some of the properties of the entities. These claims are separate from those that are derived from theories about those entities. Therefore, it is wrong to say that entity realists believe next to nothing about experimental entities.

Admittedly, the reviewer's interpretation of entity realism goes hand in hand with Hacking's contention that electron guns "rely on the causal properties of electrons" and "a modest number of home truths about electrons" (Hacking, 1983, p. 265). It also agrees with Margaret Morrison's interpretation of entity realism that the home truths about experimental entities "do not constitute anything like the kinds of complex frameworks that are normally taken to be definitive of a theory" (Morrison, 1990, p. 1). Therefore, I think that the reviewer's criticism is legitimate.

However, the reviewer's criticism is compatible with my previous contention that entity realists believe that oxygen is real, but not that the oxygen theory is true. Experimental scientists may use the home truths required to manipulate oxygen. But the home truths are not part of the oxygen theory, as the reviewer and Morrison claim. Accordingly, entity realists believe that none of the properties that the oxygen theory attributes to oxygen is real and that none of the descriptions of the oxygen theory is true. In sum, entity realists believe next to nothing regarding what the oxygen theory says about the world.

What about the causal knowledge and the home truths? We might grant that they are not part of a scientific theory. Consider, however, that we can define realism as the view that "science has achieved theoretical truth" (Psillos, 1999, p. xx). On this new definition of realism, it does not matter whether theoretical claims are part of a scientific theory. As long as you believe that they are true, your beliefs about the world go beyond what science tells us about observables, and hence you are a realist. Since entity realism claims that the theoretical claims about the causal properties of electrons are true, i.e., since entity realism asserts that science has obtained the home truths about electrons, it collapses into realism.

What if experimental scientists claim that the home truths are part of a scientific theory? Entity realists would say that we should disbelieve the home truths, given that they reject theory realism. However, it does not sound right to say that we can justifiably believe the home truths or not, depending on whether they are classified as part of a scientific theory or separate from a scientific theory, or depending on whether they constitute or do not constitute "complex frameworks that are normally taken to be definitive of a theory" (Morrison, 1990, p. 1). Why is it that the home truths are believable when they are separate from the complex frameworks, but unbelievable when they are part of the complex frameworks? Thus, it is not clear how much it helps entity realism to say that the home truths are not part of a scientific theory.

Since entity realism is committed to the home truths about experimental entities, i.e., since it claims that the causal properties of experimental entities are real, it might as well be called "home veritisim" or "causal propertism." Moreover, it opens the door to realism, as defined by Hilary Putnam (1975, p. 73), for realists would claim that justification is required for the position that causal properties are real, but noncausal properties are not, i.e., that justification is required for the double standard toward causal and noncausal properties of experimental entities. There is a further issue of whether we can acquire the home truths without using IBE. I leave these interesting issues unresolved in this book.

## 5.3 Minimal Realism

Juha Saatsi has developed an intriguing position that he calls "minimal realism." It holds that "science as a matter of fact makes *theoretical* progress in the sense that theories better supported by scientific evidence (by and large) latch better onto unobservable reality" (Saatsi, 2019, p. 3988). There is an issue of what exactly "latch better onto" means. I take it to mean that newer theories, which are better supported by evidence than older ones, are more truthlike than older ones.

Minimal realism is significantly different from realism. Realism makes the absolute claim that current theories are (approximately) true, i.e., they are true or close to the truth. By contrast, minimal realism makes the relative claim that they are closer to the truth than their precursors. There is a substantial difference between the absolute and relative theses. Two theories might be far from the truth, yet one might be more truthlike than the other (Wray, 2008, p. 323; Mizrahi, 2013, p. 401). To use an analogy, you and I might be far from Milan on a race from Rome to Milan, and yet I may be a step ahead of you (Park, 2019a, p. 151).

Saatsi takes minimal realism to be in the middle "between Stanford's neo-instrumentalism and his demanding conception of realism" (Saatsi, 2019, p. 3989). Kyle Stanford's demanding conception of realism is that "the central claims of our best scientific theories about how things stand in nature must be at least probably and/or approximately true" (Stanford, 2006, p. 6). His conception of instrumentalism is that our best scientific theories "are powerful conceptual tools for action and guides to further inquiry" (Stanford, 2006, p. 24).

Is Saatsi right about minimal realism? He would be right if realism entails minimal realism and minimal realism entails instrumentalism. Admittedly, realism entails minimal realism. However, "minimal realism does not entail instrumentalism" (Park, 2017a, p. 102). The mere fact that T is more truthlike than its precursor does not entail that it is useful. Suppose that 2% of the observational consequences of T are true and 1% of the observational consequences of its precursor are true. Note that both T and its precursor are far from empirical adequacy. In such a case, it is unlikely that T is useful. It is a relative claim that T is more truthlike than its predecessor. By contrast, it is an absolute claim that T is useful. In general, a relative thesis does not entail an absolute thesis.

Seungbae Park (2017a) develops a position called "middlism." It is intended to truly fall between realism and instrumentalism. It holds that T is more truthlike than its predecessor and that it is useful. Realism entails middlism, and middlism entails instrumentalism. Therefore, middlism is "in the middle between realism and instrumentalism" (Park, 2017a, p. 109). Middlism is compatible with the PI. Superseded theories, such as the Ptolemaic theory and the humoral theory, were useful, and their successors are also useful, and the latter are more truthlike than the former.

There is a problem with middlism. It is composed of both relative and absolute theses. The conjunction of the two disparate theses has the whiff of arbitrariness (Park, 2017a, p. 109). Why say that T is close to empirical adequacy, but T is closer to the truth than its precursor? Why not say that T is close to both the truth and empirical adequacy? It is an absolute thesis that T is close to the truth. It is also an absolute thesis that T is close to empirical adequacy. There is no whiff of arbitrariness in the conjunction of the two absolute theses. Alternatively, why not say that T is closer to both the truth and empirical adequacy than its predecessor? It is a relative thesis that T is closer to the truth than its precursor. It is also a relative thesis that T is closer to empirical adequacy than its predecessor. There is no whiff of arbitrariness in the conjunction of the two relative theses.

Saatsi claims that minimal realism has an advantage, viz., it gets around the PI. He says that "a realist attitude can be maintained in the face of the historical evidence" (Saatsi, 2019, p. 3980). The idea is that the PI refutes the position that T is true, but it does not refute the position that T is more truthlike than its precursor. Hence, minimal realism can be endorsed by those who think that the PI is correct.

However, the PI was roundly criticized in Chap. 2. The first-order PIs directed at scientists, realists, and scientific theories are all infinitely problematic. Hence, the history of science cannot be a reason for realists to retreat to minimal realism. In addition, there are six new arguments for realism, as we saw in Chap. 4. For example, the counter-induction says that "since past theories were false, present theories are true" (Park, 2018b, p. 330). These arguments indicate that minimal realism is not well motivated.

How does minimal realism relate to entity realism? It does not entail entity realism. It is one thing to say that T is more truthlike than its precursor; it is quite another to say that the theoretical entities of T exist. Even if T is more truthlike than its precursor, the theoretical entities of T might not be real. Thus, minimal realism is less committed than entity realism to the claims of science. For this reason, minimal realism does not deserve the label "realism." It is more appropriate to call it "theoretical progressivism" (Park, 2017a, p. 104).

There are disadvantages with (being close to) antirealism (Park, 2020a, p. 65). The closer a position is to antirealism, the more it is vulnerable to the criticisms that I launched against antirealism in Chap. 3. For example, antirealists are vulnerable to the PI over antirealists, as we discussed in Sect. 3.4 of Chap. 3. Minimal realism is committed so little to science that it may as well be regarded as antirealism or next to antirealism. As a result, antirealists' past philosophical mistakes constitute an inductive reason for thinking that minimal realists are making a philosophical mistake. To save space, however, I do not flesh out how minimal realism falls prey to the other criticisms against antirealism in Chap. 3.

Can minimal realism explain why science is successful? Saatsi claims that it does, saying that "science can make theoretical progress in the sense of theories' latching better and better onto reality in a way that drives theories' increasing empirical adequacy and enables them to make novel predictions" (Saatsi, 2019, p. 3990). On this account, T makes novel predictions because it latches better onto unobservables than its predecessor did.

In my view, however, minimal realism fails to explain the success of theories. Suppose that a theory was proposed for the first time in a new research field and that it has made true predictions. It is problematic "to say that it was successful because it was theoretically better than its predecessor, for it did not have a predecessor" (Park, 2017a, p. 113). In addition, a theory that latches better onto reality than its predecessor might still be far from reality. It is unlikely that such a theory can make true predictions, let alone novel ones (Park, 2017a, pp. 111–112).

## 5.4 Axiological Realism

Many prestigious philosophers of science claim that science has aims. For example, van Fraassen contends that according to scientific realism, "Science aims to give us, in its theories, a literally true story of what the world is like" (van Fraassen, 1980, p. 8), and according to constructive empiricism, "Science aims to give us theories which are empirically adequate" (van Fraassen, 1980, p. 12). Hasok Chang states that "it is perfectly sensible that some functions of the history of science serve the aims of science itself" (Chang, 2016, p. 103). Alexander Bird claims that "the constitutive aim of science is the production of knowledge" (Bird, 2019, p. 172).

In this tradition, Timothy Lyons has developed an interesting view that he calls "axiological realism." It asserts that "science seeks true theories" (Lyons, 2005, p. 167). Axiological realism can be used to explain some scientific practices. For instance, scientists choose simple theories over more complex ones because they pursue true theories (Lyons, 2005, p. 172). Another advantage of axiological realism is that it avoids Larry Laudan's objection (Laudan, 1981; Lyons, 2016a, 2017, 2018) to realism, which was explicated in Sect. 1.3 of Chap. 1. Recall that on this objection, successful but false theories from the past are counterexamples to realism. Axiological realists say that such theories are not counterexamples to axiological realism, pointing out that they do not undermine the thesis that past science sought true theories.

Is axiological realism a version of realism? There is a huge difference between the axiological thesis that science pursues true theories and the epistemic thesis that we know that science has attained true theories (Park, 2020a, p. 64). To go further, the axiological thesis also significantly differs from the instrumentalist thesis that science makes true predictions. Thus, axiological realism does not even entail that T is useful. Axiological realists believe less than instrumentalists regarding what T says about observables. Axiological realism does not deserve the label "realism."

Lyons would retort that it deserves the label because it is significantly different from axiological empiricism, which asserts that science seeks empirically adequate (EA) theories (Park, 2020a, p. 64). My response to this rejoinder is to admit that axiological realism is significantly different from axiological empiricism, but to point out that axiological realism is also significantly different from realism. Moreover, axiological realism matches up with the definition of antirealism given in Sect. 2.1 of Chap. 2. Recall that antirealism is defined as any position that opposes realism.

Let me now raise a substantive objection to axiological realism. Axiological realism is designed to get around Laudan's objection to realism. As we saw in Sect. 1.3 of Chap. 1, however, Laudan's objection is compatible with realism. Hence, Laudan's objection is not a good reason for realists to retreat to axiological realism. Moreover, there are six new arguments for realism, as we saw in Chap. 4. For example, the argument from undiscovered evidence asserts that science has attained true theories. Such arguments tell realists not to settle for the modest thesis that science pursues true theories.

There are more reasons to think that realists ought not to settle for axiological realism. It is open to the criticisms against antirealism unpacked in Chap. 3. For example, scientists might negatively respond to axiological realists, saying that "axiological realists seek true philosophical theories, but that we cannot justifiably believe that axiological realists' philosophical theories, whatever they might be, are true" (Park, 2020a, p. 66). The limits of space here prevent me from sketching how axiological realism falls prey to the other criticisms against antirealism in Chap. 3.

Finally, there is an intrinsic flaw with axiological realism, viz., it provides a teleological explanation of science (Park, 2020a, p. 70). A final cause is invoked to explain an event in a teleological explanation. In contrast to a teleological explanation, an efficient cause is invoked to explain an event in a mechanical explanation. Both types of explanations were accepted as legitimate in ancient science. In modern science, however, teleological explanations are regarded as illegitimate. Teleology is forbidden even from biology when Charles Darwin claims that species do not evolve toward a goal. It is interesting to see, however, that a teleological explanation "found its safe haven in philosophy of science, and it still flourishes in the 21st century" (Park, 2020b, p. 44).

A reviewer objects that I am too quick to dismiss teleological notions. Biologists today say that an organ performs a function in an organism and that to perform a function is to serve as a means to achieve an end. For example, a heart performs the function of pumping blood. Biologists regard a heart as "a means to the end of pumping blood" (Trommler & Hammann, 2020, p. 3). Therefore, it might be legitimate for axiological realists to talk about an aim of science.

What does it mean to say that an organ has a function? Two different answers are given by Larry Wright (1976) and Robert Cummins (1975). According to Wright, to say that an organ has a function is to say that it is an adaptation. For example, organisms with hearts did better in the battle for survival and reproduction than the ones without hearts. Thus, the notion of function can be fleshed in terms of evolutionary history. According to Cummins, by contrast, to say that an organ has a function is to say that it contributes to the maintenance of an organism. A heart pumps blood. As a result, an organism persists. The persistence of the organism, in turn, enables the heart to pump blood. Thus, the function of the heart can be cashed out in terms of efficient causation (Allen & Neal, 2020).

These two notions of function are the most prominent ones in the literature (Sober, 2018, pp. 83–89; Allen & Neal, 2020). I do not attempt to determine which of these two notions is more defensible. Instead, I observe that neither of them requires that "goal-directed systems possess some immaterial ingredient that orients them toward their appropriate end states" (Sober, 2018, p. 76). This observation has an interesting implication for axiological realists, who claim that science aims to generate true theories. They must not mean that science has an ingredient that orients it toward true theories. Then, what do they mean?

Axiological realists might answer, appealing to Wright's notion of function, that to say that science aims to generate true theories means that science performs the function of producing true theories. Just as a heart is part of an organism, so science is part of the entire human enterprise, which includes science, religion, music, art, sports, and literature. Just as a heart, by pumping blood, increased the chance that an organism survived and reproduced, so science, by producing true theories, increased the chance that the entire human enterprise perpetuates itself. In a nutshell, the function of science can be fleshed out in terms of the evolutionary history of the entire human enterprise. I leave to axiological realists the task of developing this line of thought.

Alternatively, axiological realists might suggest, appealing to Cummins's notion of function, that to say that science aims to generate true theories means that science performs the function of producing true theories. Just as a heart, by pumping blood, contributes to the survival of an organism, so science, by producing true theories, contributes to the perpetuation of the entire human enterprise. Just as the survival of an organism, in turn, enables a heart to pump blood, so the persistence of the entire human enterprise, in turn, enables science to generate true theories. In a nutshell, the function of science can be cashed out in terms of efficient causation. Again, I leave to axiological realists the task of developing this line of thought.

## 5.5 Van Fraassen's Definition

Van Fraassen's (1980) definition of realism is composed of two theses: one involves the aim of science, and the other involves the acceptance of a theory. The aim component asserts that "Science aims to give us, in its theories, a literally true story of what the world is like," and the acceptance component asserts that "acceptance of a scientific theory involves the belief that it is true" (van Fraassen, 1980, p. 8). Van Fraassen's definition of realism is utterly different from Putnam's definition of realism that this book adopts, viz., we can justifiably believe our best theories. The difference between van Fraassen's and Putnam's definitions of realism, however, is not well appreciated in the literature.

The aim component coincides with Lyons's axiological realism discussed in Sect. 5.4 of this chapter, and thus it can also be called "axiological realism" or "teleological realism." How about the acceptance component? For van Fraassen, to accept T is to make "a commitment to confront any future phenomena by means of the conceptual resources of this theory" (van Fraassen, 1980, p. 12). Thus, the acceptance component *attributes* the belief that T is true to accepters of T, those who commit to confront the world in terms of T. It follows that the acceptance component might as well be called "attributive realism." In sum, van Fraassen's definition of realism is composed of axiological realism and attributive realism.

It is clear that axiological realism is a version of putative realism, and it is distinct from epistemic realism that T is warranted, as we noted in Sect. 5.4 of this chapter. What about attributive realism? It is also a variant of putative realism, and it is totally different from epistemic realism. It attributes the belief of T to scientists, those who are committed to confronting phenomena in terms of T (Park, 2020c, p. 484). Strictly speaking, it is one thing to attribute the belief of T to scientists; it is another to hold that belief oneself, as we noted in Sect. 3.10 of Chap. 3. Attributive realism does not entail that T is true, nor does it entail that T makes true predictions.

This observation of attributive realism has the following two noteworthy implications. First, attributive realism is a form of antirealism, and thus it is subject to the criticisms that I launched against antirealism in Chap. 3. For example, attributive realists are subject to the PI over antirealists. Moreover, they have the burden of refuting the six new arguments for realism in Chap. 4.

Second, van Fraassen's following argument for constructive empiricism is problematic. He says that constructive empiricism "makes better sense of science, and of scientific activity, than realism does and does so without inflationary metaphysics" (van Fraassen, 1980, p. 73). Note that he speaks as if realism were committed to inflationary metaphysics. As we noted above, however, scientific realism, as he defines it, does not assert that T is true, and thus scientific realism is not saddled with inflationary metaphysics. It follows that what he means by realism in this argument for constructive empiricism is not the combination of axiological realism and attributive realism, but rather semantic realism that T is true.

In my view, however, it is like comparing apples and oranges for van Fraassen to compare constructive empiricism with semantic realism. Constructive empiricism is not about whether T is EA, but rather about whether science aims for EA theories, and whether the acceptance of T involves the belief that T is EA. Hence, constructive empiricism is compatible with semantic realism, and constructive empiricists have no reason to refute semantic realism. They rather have reason to refute the conjunction of axiological realism and attributive realism. Constructive empiricism and the conjunction make contradictory claims about what science aims for and about what accepters of T believe. Therefore, van Fraassen should attempt to demonstrate the superiority of constructive empiricism not over semantic realism but rather over the conjunction, i.e., his definition of scientific realism.

Van Fraassen criticizes semantic realism in another context when he gives the selectionist explanation of the success of science (van Fraassen, 1980, pp. 39–40). The target of the evolutionary explanation is not the conjunction of axiological and attributive realism, but rather semantic and epistemic realism (Park, 2020c, pp. 481–482). After all, it is intended to show that the success of science can be explained without truth. It might be a strike against semantic and epistemic realism; however, it is by no means a strike against the conjunction of axiological and attributive realism. Axiological realists can cheerfully embrace the evolutionary explanation but insist that science pursues true theories. Attributive realists can also cheerfully embrace the selectionist explanation but insist that accepters of T believe it. One may wonder why constructive empiricists give the selectionist explanation.

Van Fraassen might retort that attributive realism entails that T is true, so he is not comparing apples and oranges when he claims that constructive empiricism explains science better than scientific realism does and that the no-miracles argument is undermined by the selectionist explanation. A problem with this retort is that if attributive realism entails that T is true, attributive empiricism would also entail that T is EA. It is arbitrary to say that attributive realism asserts that T is true, but that attributive empiricism does not assert that T is EA. Now, if attributive empiricism asserts that T is EA, constructive empiricism, which embeds attributive empiricism, would fall prey to the problem of unobserved anomalies. More importantly, it would be wrong for supporters of constructive empiricism to defend constructive empiricism by saying that constructive empiricism does not assert that T is EA. As we noted in Sect. 3.10 of Chap. 3, however, if critics attack the position that T is EA, supporters of constructive empiricism typically say, "Constructive empiricists don't believe T is EA. You have committed a straw man fallacy."

Let me now turn to a debate between axiological empiricists and realists. Axiological empiricists would argue that scientific practices can be explained by the hypothesis that science aims for EA theories. By contrast, axiological realists would argue that scientific practices can be explained by the hypothesis that science aims for true theories. These explanations are all vulnerable to the charge that they are teleological. In light of this charge, axiological empiricists would argue that scientific practices can be explained by the hypothesis that science performs the function of generating EA theories. By contrast, axiological empiricists would argue that scientific practices can be explained by the hypothesis that science performs the function of generating true theories. I do not attempt to adjudicate between these rival hypotheses in this book.

How can we adjudicate between attributive empiricism and realism? The answer to this question is obvious. We should ask accepters of T whether they believe that T is true or EA. Scientists are the prominent accepters of T, and thus we should take an opinion poll of scientists. If the opinion poll finds that they believe that T is true, attributive realism wins, and attributive empiricism loses. If the opinion poll finds that they believe that T is EA, attributive empiricism wins, and attributive realism loses. There has been no debate between attributive empiricists and realists, although van Fraassen (1980) advanced these rival views about the acceptance of T more than 40 years ago. The absence of the debate is not surprising, given that the disputants of the realism debate are usually philosophers, who seldom take an opinion poll of scientists to resolve a philosophical dispute (Park, 2020c, p. 485).

It should be clear by now that both axiological and attributive realism are variants of putative realism and hence that scientific realism, as defined by van Fraassen (1980), is a form of putative realism. Ask the question "Are electrons real?" to those who fit van Fraassen's definition of realists. To your surprise, they will say, "Don't ask me whether electrons are real. Ask me, instead, whether science aims to give us true or EA theories, and whether the acceptance of T involves the belief that it is true or EA. To ask me whether electrons are real is just as inappropriate as to ask me whether moral properties are real." It is not clear to me how such cognitive agents can be called "realists." Since they are silent about their ontology, they might as well be called "antirealists."

As the discussion so far shows, van Fraassen's definition of scientific realism is far from semantic realism and epistemic realism that this book adopts. According to semantic realism, T is true. According to epistemic realism, T is warranted. It is unfortunate that the same label "scientific realism" is attached to these significantly different positions. Readers should not be misled by the label.

## 5.6 Selective Realism

### *5.6.1 An Analysis*

Selective realism asserts that we should be selective about the different parts of a scientific theory, as we have seen in Sect. 4.2 of Chap. 4. Some components are trustworthy, while others are not. The trustworthy components are those that are deployed to explain and predict events, while the untrustworthy components are those that are not. The deployed parts are called "working posits," and the undeployed parts are called "idle posits." Working posits survive scientific revolutions, whereas idle posits do not. In other words, working posits are preserved in subsequent theories, whereas idle posits are not. Accordingly, selective realism is also called "preservative realism" (Chang, 2003) and "deployment realism" (Lyons, 2016b) in the literature.

Selective realism has been an influential response to the PI. It is espoused by many venerable philosophers (Worrall, 1989; Kitcher, 1993; Psillos, 1999, 2009, 2018; Chakravartty, 2008; Enfield, 2008; Harker, 2008; Vickers, 2017). James Ladyman claims that structural realism, a variant of selective realism, "is considered by many realists and antirealists alike as the most defensible form of scientific realism" (Ladyman, 2014). Similarly, Alberto Cordero claims that selective realism is regarded as "the most promising realist project today" (Cordero, 2017, p. 245). Moti Mizrahi calls this philosophical trend a "selectivist turn" (Mizrahi, 2020, p. 25). As these philosophers' writings suggest, selective realism is in vogue today.

Selective realism is committed, while pessimism is not, to the truth of some of the theoretical claims of science. Accordingly, selective realists would claim that selective realism is a brand of realism, not a brand of putative realism. My response to this claim is to emphasize a significant difference between realism and selective realism and a significant similarity between pessimism and selective realism. The difference and the similarity support my assessment that selective realism is closer to pessimism and more removed from realism than selective realists would have us believe. Let me go over the difference and the similarity in turn.

The difference is that realism affirms, while selective realism denies, that current theories will persist without being refuted. We discussed the six new arguments for realism in Chap. 4. All of them imply that current theories will remain unfalsified. For example, the optimistic induction over realists holds that since the realists of the early twentieth century were right to believe the theories of their time, their present and future philosophical offspring are also right to believe the theories of their times. Realism asserts that at least *most* of the theoretical contents of current theories will be retained. By contrast, selective realism holds that only *some* of the theoretical contents of current theories will be retained (Park, 2019b, p. 95).

The similarity between pessimism and selective realism is that both assert that since past theories succumbed to scientific revolutions, current theories will also succumb to scientific revolutions. The only difference between pessimism and selective realism is that selective realism emphasizes, whereas pessimism does not, that some theoretical claims survive scientific revolutions. Pessimism and selective realism commonly predict that current theories will be overthrown (Park, 2017c, p. 99, 2019b, p. 94; Stanford, 2018, p. 79). Because selective realism makes this prediction, it makes a distinction between stable and unstable constituents of current theories. After all, if it were not committed to this prediction, it would be pointless to make the distinction.

What does this similarity between pessimism and selective realism show? The difference between them "is simply a difference of style or taste in applying the expression 'approximately true' rather than a substantive disagreement between them" (Stanford, 2015, p. 875). To put it another way, selective realists take it to be palatable, while pessimists take it to be unpalatable, to attribute the realist predicate to current theories, and there is no rational basis for preferring one linguistic practice over the other. Stanford's comment on the standoff between pessimists and selective realists indicates that selective realism lies between pessimism and realism, and that "selectivism" is a better name than "selective realism" or "selective pessimism." "Selective pessimism" would be just as appropriate as "selective realism" because selective realism asserts that the idle posits of current theories will be ousted (Park, 2019b, p. 94).

Unlike selective realism, realism predicts that current theories will persist and that at least most of their theoretical contents will not be falsified. Based on this prediction, realism attributes the phrase "approximately true" to current theories. Thus, there is not only a terminological dispute between pessimists and realists but also a substantive dispute between them, and the terminological dispute is grounded in the substantive dispute. In sum, it is clear that the theoretical claims of current theories to which realists are committed are thick and rich enough for realists to justifiably attribute "approximately true" to current theories, whereas it is not clear whether the theoretical claims of current theories to which selective realists are committed are thick and rich enough for them to justifiably attribute "approximately true" to current theories (Park, 2019b, p. 99).

However, selective realists would insist that they should be classified as realists on the grounds that unlike pessimists, they are committed to some theoretical claims of science and they ascribe "approximately true" to current theories. If they do, I am happy to classify them as such. However, I would classify as "super-realists" those to whom I have been referring as "realists" throughout this book. Just as selective realists wish to distance themselves from pessimists, so realists wish to distance themselves from selective realists. Super-realists point out that unlike selective realists, they disbelieve that current theories will succumb to scientific revolutions.

### 5.6.2 Common Problems

In this subsection, I summarize the problems common to both pessimism and selective realism. The common problems arise from the similarity between them mentioned in Subsection 5.6.1, viz., both assert that since past theories fell victim to scientific revolutions, current theories will also fall victim to scientific revolutions. Let me begin with the fallacy of biased statistics.

Selective realists agree with pessimists that past theories were refuted. This observation of the history of science is based upon the lists of past theories, such as Laudan's (1981, p. 33) list. As we noted in Subsection 2.4.1 of Chap. 2, however, Laudan's list includes distant past theories that were falsified before the twentieth century, but it does not include recent past theories that have remained unfalsified since the early twentieth century. It follows that like pessimism, selective realism commits the fallacy of biased statistics.

Let me move to the next common problem. Like the classic PI, selective realism overlooks the superiority of current theories over their predecessors. This superiority nullifies the pessimistic inference from past to current theories, leaving it open whether current theories will be overturned. However, selective realists believe that they will be overturned. Accordingly, they have to establish that although current theories are superior to their precursors, they will nevertheless be overthrown. Otherwise, they would be vulnerable to the charge that it is pointless to distinguish between the constituents of current theories that are presumed to survive scientific revolutions and the constituents of current theories that are presumed not to survive scientific revolutions. They would also be vulnerable to the charge that they are overly cautious about current theories.

In this context, selective realists might appeal to Stanford's (2006, p. 45) clever strategy for overcoming the objection to the classic PI that current theories are empirically more successful than their precursors. He grants the difference between past and current theories, but he argues that past and current scientists resemble one another in that they lack the cognitive capacity to think of alternatives to the theories of their times. Therefore, current theories will be overturned, just like their predecessors.

As we pointed out in Subsection 2.4.2 of Chap. 2, however, present scientists have more scientific knowledge than their predecessors, so present scientists have a greater cognitive capacity than that of their predecessors (Bird, 2007, p. 108; Park, 2011, p. 83, 2017d, p. 617; Mizrahi, 2020, p. 98). The superiority of the former over the latter makes it fallacious to think that since past scientists lacked the cognitive ability to ideate alternatives to their theories, present scientists also lack the cognitive ability to ideate alternatives to their theories. Therefore, Stanford's strategy to get around the first problem with the classic PI would not be useful to selective realists, who have the burden of showing that current theories will suffer the same fate as past theories although the former are better than the latter.

Let me turn to the next common problem. Like pessimism, selective realism violates a basic rule for assessing an argument (Park, 2019c, p. 371). To appraise an argument for p, we should engage in the details of that argument. That is, we should identify the premises for p and then determine whether they are true and whether p follows from them. We should not engage in the details of an argument for q and then claim that since the argument for q is weak, the argument for p is also weak, or that since the argument for q is strong, the argument for p is also strong. In short, it is wrong to evaluate the argument for q to accept or reject p.

Selective realists violate this basic rule. They assert that since earlier theories were approximately true, their present successors are also approximately true. This assertion indicates that because scientists' argument, say, for the phlogiston theory was good enough to show that it was approximately true, scientists' argument for the oxygen theory is also good enough to show that it is approximately true. Selective realists assess scientists' argument for the phlogiston theory, not their argument for the oxygen theory, to determine whether the oxygen theory is approximately true.

John Worrall (1989), Philip Kitcher (1993), and Stathis Psillos (1999) offer the case studies of the ether, phlogiston, and caloric theories. They observe that only the working posits of these superseded theories were worthy of our beliefs. Based on this observation, they conclude that only the working posits of their present successors are worthy of our beliefs. Their inference indicates that because scientists' argument for the phlogiston theory was good enough to show that its working constituents were true, but not that its idle ones were true, scientists' argument for the oxygen theory is also good enough to show that its working constituents are true, but not that its idle constituents are true.

However, it is a violation of the aforementioned rule to extrapolate the strength of scientists' argument for a current theory from the strength of scientists' argument for its forerunner. If you want to establish that scientists' argument for the oxygen theory is strong or weak, you should provide a case study of the oxygen theory, not a case study of the phlogiston theory. If you provide a case study of a past theory, you can make a claim about that past theory, but not about its current successor.

This criticism against selective realism complies with Peter Lipton's (2001, p. 353) insight that "at the end of the day, the only evidence for the truth of scientific theories is the evidence that scientists' use, and the only positive arguments for scientific realism are the arguments that scientists make" (Lipton, 2001, p. 353). Lipton's view implies that selective realists' observation that some theoretical assumptions of the phlogiston theory are enshrined in the oxygen theory does not support the oxygen theory at all and that pessimists' observation that the phlogiston theory was overturned does not undermine the oxygen theory in any way.

Jamin Asay suggests that the participants in the realism debate should exchange ideas with philosophers in other fields of philosophical research. If they do, they "can learn from one another" (Asay, 2019, p. 594). I agree with Asay on this count. Imagine how, say, moral philosophers would think about the selectivist methodology. When there is a series of scientific theories, selective realists look for a common theoretical constituent that is preserved in the series. Moral philosophers would regard such a method as bizarre. They do not look for a claim common to successive moral theories, such as utilitarianism and Kantianism. They rather take their attitude toward a moral theory independently of whether philosophers' arguments for its predecessor are strong or weak. They evaluate Kantians' arguments for Kantianism to determine whether it is tenable. They do not evaluate utilitarians' arguments for utilitarianism and then extrapolate the strength of Kantians' argument for Kantianism. Moral philosophers observe, while selective realists violate, the basic rule for appraising an argument (Park, 2019c, pp. 369–371).

### 5.6.3 Distinctive Problems

Selective realism has distinctive problems, problems that it does not share with pessimism. I look into them in this subsection, beginning with the problem of a criterion.

It is controversial whether we can find a viable criterion for differentiating between the stable and unstable components of successful theories, as we noted in Sect. 4.2 of Chap. 4. Without such a criterion, it is difficult to tell whether a given theoretical constituent of a present theory is trustworthy, although it is easy to tell whether a given theoretical constituent of a past theory is trustworthy. A stable posit of a past theory is one that is preserved in its present successor. A trustworthy posit of a present theory is one that will be found in its future successor. However, the future theory has not yet been conceived, and thus we do not know which constituents of a current theory are trustworthy. For this reason, selective realists attempt to make a principled distinction between the trustworthy and untrustworthy constituents of a present theory. Some philosophers (Stanford, 2006, pp. 173–180, 2009, pp. 385–387; Saatsi, 2019) are pessimistic, while other philosophers (Votsis, 2011; Vickers, 2013, 2017) are optimistic, about the prospect of making such a principled distinction.

Let me turn to another distinctive problem with selective realism. Selective realism asserts that past and present theories share certain theoretical assumptions. However, there are some historical episodes in which no common theoretical assumption was preserved in successive theories. For example, no theoretical assumption from the humoral theory was preserved in the miasma theory. If there had been selective realists in the middle of the nineteenth century, they would have believed that just as no theoretical content of the humoral theory was preserved in the miasma theory, so no theoretical content of the miasma theory would be preserved in its successor. This prediction runs counter to the historical fact that the germ theory shares with the miasma theory the theoretical assumptions that "a disease originates outside of a human body, and that diseases spread through the air" (Park, 2017c, p. 104). In sum, selective realism cannot be applied to some cases of theory-transitions in the history of science.

Selective realism would have been detrimental to the progress of science, if scientists had adopted it as their philosophical framework. Imagine that you were a scientist looking into disease at the end of the nineteenth century when the miasma theory was plagued by anomalies. If you had been a selective realist, you would have thought that just as the humoral theory was discarded in toto, so the miasma theory and its successor would be discarded in toto. As a result, you would not bother to conceive of the germ theory, thinking that it would also end up being discarded in toto. It is fortunate that germ theorists were not selective realists (Park, 2017c, p. 104).

In response to this criticism, selective realists might give up the global selective induction (SI) and embrace local SIs (Park, 2019c, p. 362). The global SI holds that a particular kind of theoretical assumptions was preserved in the theory-transitions in all fields of science, so the same kind of theoretical assumptions will survive subsequent theory-transitions in all fields of science. In contrast to the global SI, a local SI holds that a particular sort of theoretical assumptions was preserved in the theory-transitions in a particular field of science, so the same sort of theoretical assumptions will be preserved in the subsequent theory-transitions in that field. A local SI ranges over only that particular field of science. It is compatible with the fact that another sort of theoretical assumptions was preserved in another field of science, as well as with the fact that no theoretical assumption was preserved in yet another field of science.

Admittedly, a local SI is more promising than the global SI. However, a local PI shares with pessimism the prediction that a current theory will eventually be discarded. Consequently, it is liable to the objection that the current theory is superior to its predecessor as well as to the objection that it violates the basic rule for evaluating an argument (Park, 2019c, p. 371).

There is a more fundamental problem with a local SI. This form of induction starts from the reasonable assumption that the epistemic status of, say, DNA differs from that of the Higgs field, so we should evaluate scientists' beliefs about DNA and their beliefs about the Higgs field separately. Evaluating them together "is bound to ignore and erase the very real differences between the different scientific disciplines" (Asay, 2019, p. 593). However, this reasonable assumption undermines a local SI. Just as the epistemic status of DNA differs from that of the Higgs field, so the epistemic status of, say, phlogiston differs from that of oxygen. However, a local SI claims that we can evaluate the phlogiston and oxygen theories together. Specifically, it says that since the working posits of the phlogiston theory were retained, the working posits of the oxygen theory will also be retained. It also asserts that since the phlogiston theory was approximately true, the oxygen theory is also approximately true.

Contrary to what the local SI asserts, the reasonable assumption implies that we should evaluate the phlogiston and oxygen theories separately. Evaluating them together is bound to be unjust to the oxygen theory. Scientists' arguments for the oxygen theory are different from scientists' arguments for the phlogiston theory. Therefore, just as scientists' arguments concerning DNA should be evaluated independently of how scientists' arguments concerning the Higgs field are evaluated, so scientists' arguments for the oxygen theory should be evaluated independently how scientists' arguments for the phlogiston theory were evaluated (Park, 2019c, p. 366).

A reviewer observes that the phlogiston and oxygen theories were competing theories of the same domain. By contrast, the theories of DNA and the Higgs field are not competing theories of the same domain. Therefore, it is odd to draw a parallel between two contrasting cases, oxygen vs. phlogiston, on the one hand, and DNA vs. the Higgs field, on the other.

The reviewer's intuition coincides with Asay's (2019) intuition. For Asay, it is illegitimate to infer the epistemic status of the Higgs field from that of DNA, but it is legitimate to infer the epistemic status of oxygen from that of phlogiston. To put it differently, it is illegitimate to lump together DNA and the Higgs field, but it is legitimate to lump together phlogiston and oxygen. Asay's intuition differs from mine. On my intuition, it does not matter whether two theories are of the same domain or different domains. They should be evaluated separately. Just as it is wrong to lump together DNA and the Higgs field, so it is wrong to lump together phlogiston and oxygen.

Whose intuition is right? As we noted in Subsection 5.6.2 of this chapter, Asay (2019, p. 594) proposes that the participants in the realism debate should communicate with other philosophers. I welcome this proposal but point out that epistemologists would take pessimism and selective realism as bizarre. Epistemologists evaluate foudationalism, coherentism, and reliabilism separately, just as they evaluate reliabilism and Plato's theory of forms separately. They do not evaluate the three theories of justification together on the grounds that they are competing theories of the same domain (Park, 2019c, pp. 369–371).

### 5.6.4 *The PIs against Selective Realism*

In Subsections 5.6.2 and 5.6.3 of this chapter, I noted several problems with selective realism. They are summarized in the following list:

**The List of Problems with Selective Realism**
- It is controversial whether past theories were approximately true.
- Selective realism commits the fallacy of biased statistics.
- Selective realism overlooks the superiority of current theories over their predecessors.
- Selective realism violates the basic rule for assessing an argument.
- There should be a tenable distinction between working and idle posits.
- No theoretical assumption was preserved in some series of successive theories.
- Selective realism would be harmful to the progress of science.

These problems were hidden when pioneering thinkers (Worrall, 1989; Kitcher, 1993; Psillos, 1999) developed selective realism. The problems entitle us to infer that there are hitherto hidden problems with selective realism and that they will later be brought to light. Their advent will strengthen the inductive rationale for thinking that there are more such problems. It follows that selective realism has infinitely many hitherto hidden problems. These problems are not visible to us at present, not because they do not exist but rather because they are visible only to those who have the necessary cognitive capacity. We do not have the cognitive capacity, but our philosophical offspring will have the cognitive capacity. Let me call this PI "the PI against selective realism."

How would selective realists respond to the PI against selective realism? They might say that it is incorrect. On their account, selective realism is not susceptible to a PI because it is a philosophical theory, and only a scientific theory is susceptible to a PI. In my view, however, such a reply is self-serving and would invite an equally self-serving opposing position from scientists. They would say that a philosophical theory is subject to a PI while a scientific theory is not.

In addition, epistemic reciprocalists predict that selective realism will be replaced by an alternative. Either none or only a part of selective realism will be preserved in this alternative. If no part of selective realism is fated to be preserved in the alternative, then no part of it is worthy of our beliefs now. If only a part of selective realism is fated to be preserved in the alternative, then only this part is worthy of our beliefs now. Epistemic reciprocalists' attitude toward selective realism should be agreeable to selective realists because it reflects their attitude toward scientific theories.

On second thought, however, these criticisms against selective realism are supererogatory, and there is a more convenient way to show that selective realism is infinitely problematic. In Subsection 5.6.1 of this chapter, I claimed that the only difference between the classic PI and the SI is that the SI emphasizes, while the classic PI does not, that some theoretical claims are preserved. This observation of the difference between the classic PI and the SI supports the view that the SI is a successor to the classic PI, just as the kinetic theory is a successor to the caloric theory. So what? The classic PI and the SI are competing inductions of the same domain. More importantly, the SI falls victim to the grand PI, according to which since the classic PI is infinitely problematic, its successors, including the SI, are also infinitely problematic.

### *5.6.5 The Six New Arguments*

In Chap. 4, I described six new arguments for realism. Do they have positive or negative implications for selective realism? Selective realists might think that selective realism is a form of realism, so the six new arguments have positive implications for them. On close examination, however, they only have negative implications for them. Therefore, selective realists have the burden of refuting them.

The optimistic induction over realists holds that since the realists of the early twentieth century were right to believe the theories of their time, their present and future philosophical offspring are also right to believe the theories of their times. Suppose that there were realists and selective realists in the early twentieth century. They had opposing views about their best theories, such as the oxygen theory. The realists believed that most of their best theories would not succumb to scientific revolutions, while selective realists believed that most of their best theories would succumb to scientific revolutions. As a result, selective realists believed what they took to be the stable constituents of their best theories, but disbelieved what they took to be their unstable constituents. We can see from the perspective of the twenty-first century that the selective realists needlessly distinguished between the stable and unstable constituents and that they were excessively cautious about their best theories. Their philosophical offspring will suffer the same fate. Therefore, selective realism "recommends less when we deserve more" (Park, 2019b, p. 101).

The rest of the six new arguments for realism have the same implication, viz., that selective realists are overly cautious about our best theories. For example, the grand optimistic induction holds that since realism has been supported by the five previously unconceived arguments, it will continue to be supported by hitherto unconceived arguments, as we noted in Sect. 4.7 of Chap. 4. It implies that there are infinitely many unconceived arguments for realism, and they all show that it is wrong for selective realists to assert that our best theories will be overturned, and that selective realists believe less when they can believe more regarding what our best theories say about the world.

Realists can use the six new arguments in the following two different ways. First, they can challenge selective realists to refute each of the six new arguments. It would be a daunting task for selective realists to meet this challenge, given that most of the six new arguments utilize pessimists' ideas and selective realism is similar to pessimism. Second, realists can add the six new arguments to the list of the problems with selective realism in Subsection 5.6.3 of this chapter, given that they present challenges to selective realism. The expanded list would bolster the premise of the PI against selective realism. Realists can use the six new arguments in whatever way they wish, depending on their preferences.

In whatever way realists use the six new arguments, they would urge participants in the realism debate to ponder whether the selectivist turn is justifiable. Selective realism is popular these days, as mentioned in Subsection 5.6.1 of this chapter. By contrast, realism is regarded as outdated, demanding, and indefensible in light of the PI. As far as I am concerned, we have good reasons to choose realism over selective realism.

## 5.7 The Argument from a Bad Lot

As we noted in Sect. 5.3 of this chapter, Saatsi develops minimal realism to get around the PI. In addition to the PI, there are other arguments that are leveled at realism. One of them is van Fraassen's (1989) argument from a bad lot that we saw in Sect. 1.5 of Chap. 1. Many prominent philosophers of science take it to be a serious threat to realism. In this section, I assume that it is a correct argument, and then argue that it can be applied not only to realism but also to the five variants of putative realism, so it cannot be an argument for rejecting realism and accepting the five variants of putative realism.

Realists observe that T explains phenomena better than all the conceived rival theories and then infer that T is true. The argument from a bad lot holds that there might be unconceived rival theories that are better than T, i.e., T "may well be the best of a bad lot" (van Fraassen, 1989, p. 143). Therefore, realists' inference requires the prior thesis that T is better than unconceived rival theories. The target of the argument from a bad lot is realism, according to which T is true. Its premise is that T might have unconceived rival theories.

We can give a little twist to this premise so that it states that there might be unconceived theoretical entities. With this modified premise, the argument from a bad lot can be turned against entity realism. Imagine that phlogiston theorists thought that they manipulated phlogiston. It did not occur to them that they were manipulating oxygen. In this sense, they were manipulating unconceived theoretical entities. Entity realists observe that scientists manipulate oxygen and then infer that oxygen is real. The argument from a bad lot with the new premise implies that although oxygen is a better choice than phlogiston, it may be the best of a bad lot, so entity realists' inference requires the prior thesis that oxygen is a better choice than hitherto unconceived theoretical entities. In other words, entity realists have the burden of showing that their position that scientists manipulate oxygen is better than the alternative position that scientists manipulate hitherto unconceived theoretical entities.

T has semantic properties, such as (approximate) truth and (approximate) empirical adequacy. Realists believe that it has the semantic property of being (approximately) true. Empiricists believe that T has the semantic property of being EA. Relative realists (Mizrahi, 2020, p. 115) believe that T has the semantic property of being *comparatively* true, which means that T is more truthlike than its competitor. Some may assert that T is comparatively EA, which means that T is closer to empirical adequacy than its competitor, as Finnur Dellsén (2018) notes. (Approximate) truth and (approximate) empirical adequacy, comparative truth, and comparative empirical adequacy are all conceived semantic properties. I claim that T might have hitherto unconceived semantic properties.

So what? If you stake out a philosophical position invoking a particular semantic property of T, you have the burden of justifying your choice of that semantic property over unconceived semantic properties. Specifically, you have to show that the particular semantic property is a better choice than unconceived semantic properties. After all, the semantic property that you have chosen may be the best of a bad lot. For example, empiricists observe T and then infer that T is EA. They choose empirical adequacy over other semantic properties. The argument from a bad lot implies that empirical adequacy may be the best of a bad lot, so empiricists' choice requires not only the thesis that empirical adequacy is a better choice than the conceived semantic properties but also the thesis that empirical adequacy is a better choice than unconceived semantic properties. In other words, empiricists' inference requires not only the thesis that empiricism is better than other conceived philosophical positions invoking the other conceived semantic properties, such as comparative empirical adequacy, but also the thesis that empiricism is better than unconceived philosophical positions invoking unconceived semantic properties. It is wrong to think that only realists have the burden of showing that truth is a better choice than other semantic properties.

The argument from a bad lot can also be directed at minimal realists. Minimal realists observe T and then infer that T is more truthlike than its forerunner. To use Mizrahi's (2020, p. 115) technical term, minimal realists infer that T is *comparatively* true. The argument from a bad lot implies that comparative truth may be the best of a bad lot, so minimal realists' inference requires not only the thesis that comparative truth is a better choice than the conceived semantic properties, such as comparative empirical adequacy, but also the thesis that comparative truth is a better choice than unconceived semantic properties. In other words, minimal realists' inference requires not only the thesis that minimal realism is better than other conceived philosophical positions invoking the other conceived semantic properties, such as comparative empirical adequacy, but also the thesis that minimal realism is better than unconceived philosophical positions invoking unconceived semantic properties.

The argument from a bad lot also works against axiological realists, attributive realists, and selective realists—all of whom choose truth over other semantic properties to develop their philosophical positions. They observe science and then infer, respectively, that *truth* is an aim of science, that accepters of T believe that T is *true*, and that the stable components of T are *true*. The argument from a bad lot implies that truth may be the best of a bad lot, so their choice of truth requires not only the thesis that truth is a better choice than the conceived semantic properties but also the thesis that truth is a better choice than unconceived semantic properties. In other words, their inferences require not only the thesis that axiological, attributive, and selective realism are better than other conceived rival positions invoking the other conceived semantic properties but also the thesis that axiological, attributive, and selective realism are better than unconceived rival positions invoking unconceived semantic properties.

The five versions of putative realism are close to antirealism, as we saw in the previous section. As we have just seen in this section, however, they do not fare better than realism vis-à-vis the argument from a bad lot. Therefore, the argument from a bad lot cannot be the ground for philosophers to retreat from realism to the five forms of putative realism. Alternatively, those who are convinced of the argument from a bad lot and retreat from realism to the five variants of putative realism have the burden of showing that the argument from a bad lot refutes realism, but not the five variants of putative realism.

## 5.8 Summary

Fine (1986) claims that there are too many realisms. This chapter shows, I hope, that there is only one realism and that all of the other views that use the name "realism" are instead versions of putative realism. The variants of putative realism make diverse assertions about science. However, their assertions are so meager and thin that they do not deserve the name "realism." Those who begin to take up the realism debate should not be misled by the name.

Admittedly, selective realism is the closest to realism among the variants of putative realism in the literature. However, there is a significant difference between realism and selective realism, and there is a significant resemblance between pessimism and selective realism. Pessimists and selective realists have to form an alliance against realists to establish that current theories will eventually be abandoned. For this reason, I classify selective realism as a form of putative realism, contrary to its popular conception, and to the dismay of selective realists. Selective realism faces both the PI and the grand PI against it. Finally, the argument from a bad lot, if correct, demolishes not only realism but also the five variants of putative realism.

## References

Alai, M. (2017). Resisting the historical objections to realism: Is Doppelt's a viable solution? *Synthese, 194*(9), 3267–3290.

Allen, C., & Neal, J. (2020). Teleological notions in biology. In E. N. Zalta (Ed.) *The Stanford encyclopedia of philosophy*. URL = https://plato.stanford.edu/archives/spr2020/entries/teleology-biology/

Asay, J. (2019). Going local: A defense of methodological localism about scientific realism. *Synthese, 196*(2), 587–609.

Bird, A. (2007). What is scientific progress? *Noûs, 41*(1), 92–117.

Bird, A. (2019). The aim of belief and the aim of science. *Theoria: International Journal for Theory, History and Foundations of Science, 34*(2), 171–193.

Chakravartty, A. (2008). What you don't know can't hurt you: Realism and the unconceived. *Philosophical Studies, 137*(1), 149–158.

Chakravartty, A. (2017). Scientific realism. In E. N. Zalta (Ed.) *The Stanford encyclopedia of philosophy*. URL = https://plato.stanford.edu/archives/sum2017/entries/scientific-realism/

Chang, H. (2003). Preservative realism and its discontent: Revisiting caloric. *Philosophy of Science, 70*(5), 902–912.

Chang, H. (2016). Who cares about the history of science? *The Royal Society Journal of the History of Science, 71*(1), 91–107.

Cordero, A. (2017). Retention, truth-content and selective realism. In E. Agazzi (Ed.), *Varieties of scientific realism: Objectivity and truth in science* (pp. 245–256). Springer.

Cummins, R. (1975). Functional analysis. *The Journal of Philosophy, 72*(20), 741–765.

Dellsén, F. (2018). Scientific progress: Four accounts. *Philosophy Compass, 13*(11). https://doi.org/10.1111/phc3.12525

Enfield, P. (2008). Review. Exceeding our grasp: Science, history, and the problem of unconceived alternatives. K. Stanford. *The British Journal for the Philosophy of Science, 59*(4), 881–895.

Fine, A. (1986). Unnatural attitudes: Realist and instrumentalist attachments to science. *Mind, 95*(378), 149–179.

Hacking, I. (1983). *Representing and intervening: Introductory topics in the philosophy of science*. Cambridge University Press.

Harker, D. (2008). P. Kyle Stanford, Exceeding our grasp: Science, history, and the problem of unconceived alternatives. *Philosophy of Science, 75*(2), 251–253.

Kitcher, P. (1993). *The advancement of science: Science without legend, objectivity without illusions*. Oxford University Press.

Ladyman, J. (2014). Structural realism. In E. N. Zalta (Ed.), *The Stanford encyclopedia of philosophy*. URL = http://plato.stanford.edu/archives/spr2014/entries/structural-realism/

Lipton, P. (2001). Quests of a realist. *Metascience, 10*(3), 347–353.

Laudan, L. (1981). A confutation of convergent realism. *Philosophy of Science, 48*(1), 19–49.

Lyons, T. (2005). Toward a purely axiological scientific realism. *Erkenntnis, 63*(2), 167–204.

Lyons, T. (2016a). Scientific realism. In P. Humphreys (Ed.), *The Oxford handbook of philosophy of science* (pp. 564–584). Oxford University Press.

Lyons, T. (2016b). Structural realism versus deployment realism: A comparative evaluation. *Studies in History and Philosophy of Science Part A, 59*, 95–105.

Lyons, T. (2017). Epistemic selectivity, historical threats, and the non-epistemic tenets of scientific realism. *Synthese, 194*(9), 3203–3219.

Lyons, T. (2018). Four challenges to epistemic scientific realism. *Spontaneous Generations: A Journal for the History and Philosophy of Science, 9*(1), 146–150.

Lyons, T. (2019). Systematicity theory meets Socratic scientific realism: The systematic quest for truth. *Synthese, 196*(3), 833–861.

Mizrahi, M. (2013). The argument from underconsideration and relative realism. *International Studies in the Philosophy of Science, 27*(4), 393–407.

Mizrahi, M. (2020). *The relativity of theory: Key positions and arguments in the contemporary scientific realism/antirealism debate*. Springer.

Morrison, M. (1990). Theory, intervention and realism. *Synthese, 82*(1), 1–22.

Müller, F. (2015). The pessimistic meta-induction: Obsolete through scientific progress? *International Studies in the Philosophy of Science, 29*(4), 393–412.

Musgrave, A. (2017). Strict empiricism versus explanation in science. In E. Agazzi (Ed.), *Varieties of scientific realism: Objectivity and truth in science* (pp. 71–93). Springer.

Park, S. (2011). A confutation of the pessimistic induction. *Journal for General Philosophy of Science, 42*(1), 75–84.

Park, S. (2017a). Critiques of minimal realism. *PRO, 92*, 102–114.

Park, S. (2017b). Defense of epistemic reciprocalism. *Filosofija. Sociologija, 28*(1), 56–64.

Park, S. (2017c). Selective realism *versus* individual realism for scientific creativity. *Creativity Studies, 10*(1), 97–107.

Park, S. (2017d). Why should we be pessimistic about antirealists and pessimists? *Foundations of Science, 22*(3), 613–625.

Park, S. (2018a). The pessimistic induction and the Golden Rule. *PRO, 93*, 70–80.

Park, S. (2018b). The anti-induction for scientific realism. *Grazer Philosophische Studien, 95*(3), 329–342.

Park, S. (2019a). Should scientists embrace scientific realism or antirealism? *Philosophical Forum, 50*(1), 147–158.

Park, S. (2019b). Optimistic realism over selectivism. *Kriterion: Journal of Philosophy, 31*(1), 89–106.

Park, S. (2019c). Localism vs. individualism for the scientific realism debate. *Philosophical Papers, 48*(3), 359–377.

Park, S. (2020a). Critiques of axiological realism and surrealism. *Acta Analytica, 35*(1), 61–74.

Park, S. (2020b). Replies to Healey's comments regarding van Fraassen's positions. *Social Epistemology Review and Reply Collective, 9*(1), 38–47.

Park, S. (2020c). Formulational vs. epistemological debates concerning scientific realism. *Dialogue: Canadian Philosophical Review, 59*(3), 479–496.

Psillos, S. (1999). *Scientific realism: How science tracks truth*. Routledge.

Psillos, S. (2009). Grasping at realist straws. *Metascience, 18*(3), 363–370.

Psillos, S. (2018). Tolstoy's argument: Realism and the history of science. *Spontaneous Generations: A Journal for the History and Philosophy of Science, 9*(1), 68–76.

Putnam, H. (1975). *Mathematics, matter and method: Philosophical papers, volume 1*. Cambridge University Press.

Saatsi, J. (2019). Historical inductions, old and new. *Synthese, 196*(10), 3979–3993.

Sober, E. (2018). *Philosophy of biology*. Westview Press.

Stanford, K. (2006). *Exceeding our grasp: Science, history, and the problem of unconceived alternatives*. Oxford University Press.

Stanford, K. (2009). Grasping at realist straws. *Metascience, 18*(3), 379–390.

Stanford, K. (2015). Catastrophism, uniformitarianism, and a scientific realism debate that makes a difference. *Philosophy of Science, 82*(5), 867–878.

Stanford, K. (2018). A fond farewell to 'approximate truth'? *Spontaneous Generations: A Journal for the History and Philosophy of Science, 9*(1), 78–81.

Trommler, F., & Hammann, M. (2020). The relationship between biological function and teleology: Implications for biology education. *Evolution: Education and Outreach, 13*(11). https://doi.org/10.1186/s12052-020-00122-y

van Fraassen, B. (1980). *The scientific image*. Oxford University Press.

van Fraassen, B. (1989). *Laws and symmetry*. Oxford University Press.

Vickers, P. (2013). A confrontation of convergent realism. *Philosophy of Science, 80*(2), 189–211.

Vickers, P. (2017). Understanding the selective realist defence against the PMI. *Synthese, 194*(9), 3221–3232.

Votsis, I. (2011). The prospective stance in realism. *Philosophy of Science, 78*(5), 1223–1234.

Worrall, J. (1989). Structural realism: The best of both worlds. *Dialectica, 43*(1–2), 99–124.
Wray, B. (2008). The argument from underconsideration as grounds for anti-realism: A defence. *International Studies in the Philosophy of Science, 22*(3), 317–326.
Wright, L. (1976). *Teleological explanations: An etiological analysis of goals and functions*. University of California Press.

# Chapter 6
# Formulating Scientific Realism and Antirealism

**Abstract** Putnam's formulation of realism and antirealism can trigger both *formulational* and *epistemological* debates. By contrast, van Fraassen's formulation of realism and empiricism can trigger formulational debates, but not epistemological ones. Realism is significantly different from antirealism under Putnam's formulation, while realism does not considerably differ from empiricism under van Fraassen's. Hence, there can be more debates under Putnam's formulation than under van Fraassen's. The epistemic formulation, the attributive formulation, and the conditional formulation concern different aspects of science. Indispensabilists in philosophy of mathematics can make use of the epistemic formulation, but not the attributive and conditional formulations. The conditional formulation has both a conceptual problem and an empirical problem. Doxastic conditionalism is better than both conditional realism and conditional empiricism. Finally, there is a disadvantage to formulating realism and antirealism in terms of "acceptance."

**Keywords** Attributive formulation · Conditional formulation · Doxastic conditionalism · Epistemic formulation · Epistemological debate · Formulational debate

## 6.1 Introduction

Writers formulate realism and antirealism differently. Some formulations generate voluminous debates between realists and antirealists, while others do not. What makes the former fertile and the latter less so? What is the distinctive feature of the former? There are different kinds of formulations of realism and antirealism in the literature, and they concern different aspects of science. Which aspects of science do they attempt to reveal? This chapter aims to answer these questions. My answers to them will hopefully cast light on what sorts of debates realists and antirealists are engaged in, on how realism and antirealism should be formulated, and on which aspects of science the discussants in the realism debate attempt to reveal.



S. Park, *Embracing Scientific Realism*, Synthese Library 445,
https://doi.org/10.1007/978-3-030-87813-9_6

The outline of this chapter is as follows. In Sect. 6.2, I argue that Hilary Putnam's (1975) formulation can stimulate both what I call "formulational debates" and "epistemological debates." By contrast, Bas van Fraassen's (1980) formulation can lead to formulational debates, but not to epistemological ones. In Sect. 6.3, I argue that realism significantly differs from antirealism under Putnam's (1975) formulation, while realism does not significantly differ from antirealism under van Fraassen's (2017) formulation; hence, there can be more debates under Putnam's formulation than under van Fraassen's formulation. In Sect. 6.4, I compare what I call "the epistemic formulation" (Putnam, 1975), "the attributive formulation" (van Fraassen, 1980), and "the conditional formulation" (Dellsén, 2017). These three kinds of formulations attempt to uncover different aspects of science. In Sect. 6.5, I argue that indispensabilists in philosophy of mathematics can make use of the epistemic formulation, but not the attributive or conditional formulations. In Sect. 6.6, I raise conceptual and empirical objections to both conditional realism and empiricism. In Sect. 6.7, I propose an alternative to conditional realism and empiricism. In Sect. 6.8, I attempt to show that there is a disadvantage to formulating realism and antirealism in terms of "acceptance." This criticism applies to both van Fraassen's (1980) formulation and Finnur Dellsén's (2017) formulation.

## 6.2 Formulational vs. Epistemological

In this section, I distinguish between formulational and epistemological debates (Park, 2020a, pp. 480–482). A formulational debate concerns how realism and antirealism should be defined. By contrast, an epistemological debate concerns whether there is enough evidence for a position defined in a particular manner. This distinction is useful for evaluating the diverse formulations of rival positions in the literature.

Formulations of rival positions that can stimulate both formulational and epistemological debates can yield more insights into science, and hence they can be more useful to philosophers than those that can stimulate formulational debates, but not epistemological ones. Putnam's (1975) formulation can engender both kinds of debates. By contrast, van Fraassen's (1980) formulation can engender formulational debates, but not epistemological ones. Consequently, Putnam's formulation can be more useful to philosophers than van Fraassen's (Park, 2020a, pp. 487–488). Let me flesh out this argument in what follows.

Under Putnam's (1975, p. 73) formulation, realism affirms, while antirealism denies, that successful theories are true. As we saw in Chaps. 1 and 2, Putnam constructs the no-miracles argument (NMA) to justify realism, and pessimists construct the pessimistic induction (PI) to justify antirealism. Realists and pessimists have opposing views about whether we have sufficient evidence for scientific theories. Consequently, we can conclude that Putnam's formulation induced epistemological debates.

In response to the PI, some realists propose enhanced versions of realism. My version is that successful theories that cohere with each other are true (Park, 2011). Alan Musgrave (1985) and Jarrett Leplin (1997) propose that theories demonstrating novel success are true. In response, Timothy Lyons (2003, pp. 898–899; 2017, p. 3204) and Peter Vickers (2017, p. 3227) present historical counterexamples of obsolete theories that demonstrated novel success. By contrast, selective realists argue that the stable assumptions of past theories were trustworthy, so the stable assumptions of current theories are trustworthy. In addition to selective realism, there are other forms of putative realism, as we noted in Chap. 5. In sum, the PI led some theorists to raise the bar for believable theories and others to retreat from realism to putative realism. Therefore, we can conclude that Putnam's formulation of realism has indeed induced formulational debates.

Putnam's formulation of realism radically differs from van Fraassen's. According to the latter, scientific realism is the position that "Science aims to give us, in its theories, a literally true story of what the world is like," and that "acceptance of a scientific theory involves the belief that it is true" (van Fraassen, 1980, p. 8). Constructive empiricism is the position that "Science aims to give us theories which are empirically adequate," and "acceptance of a theory involves as belief only that it is empirically adequate" (van Fraassen, 1980, p. 12). Van Fraassen's formulation invokes the aims of science and the technical notion of acceptance.

Richard Healey constructs an alternative to constructive empiricism: "Science aims to give us theories some of whose observational consequences are true: and acceptance of a theory involves as belief only that some of its observational consequences are true" (Healey, 2019b, p. 28). This alternative is less ambitious than constructive empiricism, just as constructive empiricism is less ambitious than scientific realism as defined by van Fraassen. However, it is controversial whether this alternative is legitimate. Seungbae Park (2019a, 2020b) argues that it is, whereas Healey (2019a, b) argues that it is not. Note that van Fraassen's formulation has produced a formulational debate between Healey and me.

The aim components of scientific realism and constructive empiricism have stimulated a formulational debate. Evolutionary theory denies that natural selection is a goal-oriented process, and Thomas Kuhn argues that neither organisms nor science move toward a goal, as we saw in Sect. 3.2 of Chap. 2. Van Fraassen gives an evolutionary explanation for the success of science, as we saw in Sect. 1.5 of Chap. 1. However, he defines scientific realism and constructive empiricism in terms of the teleological outlook that science has aims. As a result, Kuhn would object that if van Fraassen were to take evolutionary theory seriously, he would not formulate the two rival positions in terms of the aims of science (Park, 2020a, pp. 482–483). Note that there is and can be a debate over whether it is acceptable to formulate rival positions in terms of the aims of science.

As we saw in Sect. 5.5 of Chap. 5, the acceptance component of scientific realism attributes to accepters of T, those who have made the commitment to using T for scientific purposes, the belief that T is true, while the acceptance component of constructive empiricism attributes to accepters of T the belief that T is empirically adequate (EA). I called the acceptance components of the two rival positions "attributive realism" and "attributive empiricism," respectively. Attributive realism and empiricism attribute different beliefs to accepters of T, and they are rival descriptions of what accepters of T believe regarding T.

We should not conflate attributive realism and empiricism with epistemic realism and empiricism (Park, 2020a, pp. 484–487). Epistemic realism and empiricism are the rival views, respectively, that we have sufficient evidence for the belief that T is true and for the belief that T is EA. By contrast, attributive realism and empiricism are the rival views, respectively, that accepters of T believe that T is true and EA. Epistemic realism and empiricism concern how much evidence we have for T, whereas attributive realism and empiricism concern what accepters of T believe. To reiterate, attributive realism and empiricism do not claim that we can justifiably believe that T is true and EA, respectively.

Relatedly, to establish that we have sufficient evidence for the belief that T is true is not to refute attributive empiricism. Suppose that the NMA is correct, such that we can justifiably believe T. Even so, attributive empiricism might still be true, i.e., accepters of T might believe that T is merely EA. In addition, to establish that we have sufficient evidence only for believing that T is EA is not to refute attributive realism. Suppose that the problem of underdetermination makes it only permissible to believe that T is EA. Even so, attributive realism might still be true, i.e., accepters of T might believe that T is true.

The history of science does not spell trouble for attributive realism and empiricism. Admittedly, past theories were false and not EA, as we noted in Sect. 3.4 of Chap. 3. However, it is an entirely separate issue whether past scientists believed that their theories were false and not EA. It might be the case that they believed that their theories were true and EA, although their theories were false and not EA. The refutation of attributive realism and empiricism does not require the thesis that past theories were false and not EA, but rather the thesis that past scientists believed that their theories were false and not EA. Therefore, the PI, even if cogent, does not refute attributive realism and empiricism (Park, 2020a, p. 485).

How should we adjudicate between attributive realism and empiricism? We should conduct an opinion poll of scientists, as we discussed in Sect. 5.5 of Chap. 5. If the opinion poll shows that they believe that T is true, attributive realism would turn out to be true. If it shows that they believe that T is EA, attributive empiricism would turn out to be true. It is wrong to run the NMA to justify attributive realism and to run the problem of underdetermination to justify attributive empiricism. It is also wrong to advance the evolutionary explanation for the success of science to refute attributive realism. The evolutionary explanation may be a strike against epistemic realism, but not against attributive realism.

Let me elucidate the implication of my contention that attributive realism and empiricism are not epistemic theses concerning how much evidence we have for T, but rather descriptive theses concerning what accepters of T believe regarding T. Attributive realism and empiricism cannot engender an epistemological debate between rival philosophers concerning whether we are justified in believing that T is true or EA. They can spark, though, a psychological debate between rival psychologists over what accepters of T believe regarding T (Park, 2020a, p. 485). Of course, this critical comment against attributive realism and empiricism does not mean that they are worthless positions; it only means that they are different in kind from epistemic realism and empiricism.

A formulation of realism and antirealism is a proposal for how to use the words "realism" and "antirealism." Consequently, the formulation is either useful or useless, and it is neither true nor false. The more a formulation of realism and antirealism generates debates, the more insights it yields about science, and hence the more useful it is. Therefore, we cannot claim that Putnam's formulation is true, while van Fraassen's formulation is false. We can claim at most that the former is more useful than the latter (Park, 2020a, pp. 489–490).

A reviewer raises the following worry. It is not by any means obvious why I claim that van Fraassen's constructive empiricism does not give rise to epistemological debates. The question of whether there is good reason to believe in the truth rather than the empirical adequacy of a theory is precisely an epistemological question. The whole point of the challenge that constructive empiricism raises against scientific realism is an epistemological point. To respond to constructive empiricism, the scientific realist must engage in some epistemology.

The reviewer's worry is shared by many philosophers who interpret constructive empiricism as advancing the epistemic thesis that we can justifiably believe that T is EA, but not that it is true. Unfortunately, I could not find textual evidence for their interpretation of constructive empiricism in van Fraassen's (1980) definition of constructive empiricism. Moreover, when van Fraassen's opponents criticize the position that T is EA, his proponents standardly retort that he does not believe that T is EA, as we noted in Sect. 3.10 of Chap. 3. I found it hard to reconcile the standard retort with the epistemic thesis. For these reasons, I disagree with the reviewer on whether constructive empiricism poses an epistemological challenge to epistemic realism.

But I admit that constructive empiricism gave rise to epistemological debates in the following sense. Some discussants of the realism debate erroneously superimposed an epistemic thesis on constructive empiricism. For example, David Papineau interprets constructive empiricism as claiming that "we ought never to believe in the truth of any theory which goes beyond the observable phenomena" (Papineau, 1996, p. 8). Thanks to such misinterpretations, philosophers were able to engage in epistemological debates over whether we can justifiably believe what science tells us about unobservables. Consequently, we can say that constructive empiricism indirectly gave rise to epistemological debates.

The right interlocutor for epistemic realists is not van Fraassen but Brad Wray, one of the leading figures in the contemporary realism debate. He puts forward the epistemic thesis that "all that scientists can claim to know about their background theories is that they save the phenomena" (Wray, 2008, p. 321). This epistemic thesis can generate epistemological debates between epistemic realists and empiricists. Since Wray states that T is EA, his defenders would not say to his critics, "Wray doesn't believe T is EA. So you committed a straw man fallacy!" I do not rule out the possibility, though, that they might play the gap card. In other words, they might say, "There is a gap between what Wray says and what he believes. So your criticism doesn't refute what he believes!"

The reviewer might raise the following objection. How about the evolutionary explanation (van Fraassen, 1980, pp. 39–40)? Van Fraassen proposes it to undermine the NMA, saying that the success of science can be explained without truth, as we discussed in Sect. 1.6 of Chap. 1. He clearly engages in an epistemological debate with epistemic realists over whether the NMA is correct. Therefore, he generates epistemological debates.

Strictly speaking, however, constructive empiricism does not require the refutation of the NMA because constructive empiricism and epistemic realism are different sorts of theses and they are compatible with each other, as we discussed in Sect. 5.5 of Chap. 5. It is one thing that the success of science can be explained without truth; it is another that science aims to generate EA theories and accepters of T believe that it is EA. The evolutionary explanation is not part of constructive empiricism any more than van Fraassen's belief in God is part of constructive empiricism. Therefore, it is one thing that the evolutionary explanation generates epistemological debates; it is another that constructive empiricism generates epistemological debates.

## 6.3 Wide and Narrow Formulations

The diverse formulations of realism and antirealism in the literature can be divided into wide and narrow formulations, depending on whether there is a substantive difference between the rival positions (Park, 2019b, pp. 483–485). In a wide formulation, there is a substantive difference, whereas in a narrow formulation, there is no such difference. Under a wide formulation, there can be copious debates between rival parties. Under a narrow formulation, by contrast, there can be no such debate. Consequently, a wide formulation is preferable to a narrow one.

Which formulations in the literature are wide? Putnam's (1975, p. 73) formulation is a wide one (Park, 2019b, p. 484). Under his formulation, realism affirms, while antirealism denies, that successful theories are true. The gap between these rival positions is so vast that many philosophers have staked out positions that appear to fall between realism and antirealism and that there have been voluminous debates among realists, putative realists, and antirealists in the literature. In fact, realists defend the NMA, while pessimists defend the PI, to justify their respective positions. These two arguments have laid the ground for the realism debate over the past several decades, as stated in Chaps. 1 and 2.

Let me give an example of a narrow formulation (van Fraassen, 1989, pp. 171–172; 2017). Suppose that realists believe T but antirealists do not believe T. Under the English notion of rationality, however, realists concede that antirealists' disbelief of T is reasonable, and antirealists concede that realists' belief of T is reasonable. To put it differently, realists and antirealists agree that it is rational for the former to believe and for the latter to disbelieve T, as we noted in Sect. 4.6 of

Chap. 4. This formulation of the rival positions is so narrow that opposing parties do not have much to debate. When they meet with each other for a debate about science, they can talk about the weather and go back home.

Under the English notion of rationality, it is difficult for realists to refute antirealism, not because antirealism is justified by powerful arguments, but rather because it does not substantially differ from realism; and it is difficult for antirealists to refute realism, not because realism is justified by powerful arguments, but rather because it does not substantially differ from antirealism. Antirealists' request for realists to refute antirealism nearly amounts to the request that realists refute realism, and realists' request that antirealists refute realism nearly amounts to the request that antirealists refute antirealism (Park, 2019b, p. 485).

Van Fraassen (1989, 2017) operates under the English notion of rationality, but other participants of the realism debate have been operating under the Prussian notion of rationality. It holds that "what is rational to believe is exactly what one is rationally compelled to believe" (van Fraassen, 1989, p. 171). How does van Fraassen's position relate to the positions staked out by the other participants? What are the similarities and differences between them? Is there a wide gap or a narrow gap between them?

Let me compare van Fraassen's position with Musgrave's (2017) position. Van Fraassen disbelieves T and contends that it is rational to believe T and also rational to disbelieve T. By contrast, Musgrave believes T and claims that "*it is reasonable to believe that* H is true" (Musgrave, 2017, p. 80). He also argues that "truth explains empirical adequacy better than empirical adequacy does, because the latter 'explanation' is completely circular" (Musgrave, 2017, p. 87). Consequently, it is irrational to disbelieve T. Van Fraassen accuses Musgrave of being reticent about "what explanation is good for, or what purposes it may serve" (van Fraassen, 2017, p. 101). In sum, van Fraassen and Musgrave agree on whether it is rational to believe T. However, they disagree on whether it is rational to disbelieve T. Van Fraassen claims that it is rational to disbelieve T, whereas Musgrave claims that it is irrational to disbelieve T. Therefore, there can be a debate between them over whether it is rational to disbelieve T. If Musgrave shows that inference to the best explanation is compelling, he would win the debate.

Let me now compare van Fraassen's position with the position of pessimists, such as Wray (2013) and Thomas Nickles (2017). Pessimists believe that T is false, so they would say that it is rational to disbelieve T and irrational to believe T. Accordingly, van Fraassen and pessimists would agree on whether it is rational to disbelieve T. However, they would disagree on whether it is rational to believe T. Van Fraassen claims that it is rational to believe T, whereas pessimists would claim that it is irrational to believe T. Therefore, there can be an interesting debate between them over whether it is rational to believe T (Park, 2020c, pp. 4–5). Van Fraassen is welcome to make use of the criticisms that I raised against the PI in Chap. 2 to show that it is rational to believe T.

The PI is not the only objection that antirealists have raised against realism. They have proposed antirealist explanations, alternatives to the NMA, as we saw in Sect. 3.2 of Chap. 3. Van Fraassen's selectionist explanation is one of them. They have also constructed the argument from a bad lot and raised the problem of underdetermination. The authors of all these objections would claim that the belief of T is unreasonable due to their objections to it. However, van Fraassen claims that the belief of T is reasonable, appealing to the English notion of rationality. The gap between these opposing positions is wide enough to generate an interesting debate between the rivaling debaters (Park, 2020c, p. 5).

## 6.4 Epistemic, Attributive, and Conditional Formulations

This section clarifies three kinds of formulations of realism and empiricism that I call "the epistemic formulation," "the attributive formulation," and "the conditional formulation." I argue that these three different formulations attempt to reveal different aspects of science.

Under the epistemic formulation, epistemic realism contends that we have sufficient *evidence* for believing that T is true, and epistemic empiricism contends that we have sufficient evidence for believing that T is EA, but not for believing that T is true. Accordingly, these rival positions are named "epistemic realism" and "epistemic empiricism," respectively. The epistemic formulation is inferred from Nickles's definition of realism, which says that we "have sufficient evidence and argument to conclude with confidence that even our most mature theories are true, or very nearly, that at best minor tweaking will be necessary" (Nickles, 2017, p. 151). Epistemic realism and empiricism are rival theses about how much evidence we have for T.

Under the attributive formulation, attributive realism asserts that "acceptance of a scientific theory involves the belief that it is true" (van Fraassen, 1980, p. 8), and attributive empiricism asserts that "acceptance of a theory involves as belief only that it is empirically adequate" (van Fraassen, 1980, p. 12). These rival positions *attribute* different beliefs to accepters of T. Attributive realism attributes to them the belief that T is true, while attributive empiricism attributes to them the belief that T is EA (Park, 2020a, p. 484). Accordingly, these rival positions are named "attributive realism" and "attributive empiricism," respectively.

Under the conditional formulation (Dellsén, 2017), conditional realism holds that scientists can accept T, provided that they have evidence for believing that T is true, and conditional empiricism holds that scientists can accept T, provided that they have evidence for believing that T is EA. These rival positions propose different *conditions* under which scientists can accept T. Consequently, these rival positions are named "conditional realism" and "conditional empiricism," respectively. The conditional formulation is advanced by Dellsén (2017). He defines conditional realism as the view that a "theory T should only be accepted in a scientific context if T's truth is supported by the available evidence" (Dellsén, 2017, p. 104). He defines conditional empiricism as the view that a "theory *T* should only be accepted in a scientific context if *T*'s empirical adequacy is supported by the available evidence" (Dellsén, 2017, p. 106).[1]

For Dellsén, to accept a theory is to adopt the policy of using "the theory for various scientific purposes, e.g., as the basis for empirical predictions, as premises in one's explanations, and as a guide in further theory construction" (Dellsén, 2017, p. 99). Let me give an example to illustrate the notion of acceptance. Quantum mechanics makes bizarre claims about microscopic objects. We cannot even visualize the state of affairs that it describes. Thus, it is difficult to *believe* quantum mechanics. However, quantum mechanics makes true predictions, and thus we may *accept* it. To say so means roughly that we may decide to use it to predict events, although we do not believe that it is true.

As this example of quantum mechanics indicates, the distinction between acceptance and belief is based upon the antirealist idea that it is one thing to decide to use T for scientific purposes, and it is another to believe that T is true. Antirealists would say that accepters of quantum mechanics may believe that it is false, EA, and useful. In other words, you can decide to use quantum mechanics for scientific purposes, although you do not believe that it is true and you instead believe that it is false, EA, and useful. However, if you believe that it is useless, it would be irrational for you to decide to use it for scientific purposes. Therefore, the belief that it is useless is not available to accepters of it.

Let me turn back to conditional realism and empiricism. According to conditional realism, scientists can accept T, i.e., they can decide to use T for scientific purposes, provided that they have evidence for believing that T is true. According to conditional empiricism, by contrast, scientists can accept T, i.e., they can decide to explain and predict events in terms of T, provided that they have evidence for believing that T is EA. Thus, conditional realism places a heavier epistemic requirement than conditional empiricism on accepters of T.

What if scientists have no evidence for believing that T is EA? They would then have no evidence for the truth of T. Therefore, both conditional realists and empiricists would say that scientists should not accept T. What if scientists have evidence for believing that T is EA, but not for believing that T is true? Conditional empiricists would say that it is permissible, whereas conditional realists would say that it is impermissible, for scientists to decide to use T for scientific purposes. What if scientists have evidence for believing that T is true? They a fortiori have evidence for believing that T is EA. Hence, both conditional realists and empiricists would say that scientists may accept T. However, they would still disagree about how much evidence is required to accept T. Conditional realists would insist, whereas conditional empiricists would not, that evidence for the truth of T is required to accept T.

How does Dellsén's conditional formulation differ from van Fraassen's attributive formulation? The attributive formulation presents rival *descriptive* theses about what accepters of T believe regarding T, whereas the conditional formulation

---

[1] Dellsén (2017) uses "reconstructed realism" and "reconstructed empiricism" instead of "conditional realism" and "conditional empiricism." It seems to me, however, that "conditional" better characterizes his formulation than "reconstructed."

presents rival *normative* theses about under what epistemic conditions scientists can accept T. In other words, the attributive formulation concerns what scientists believe when they decide to use T for scientific purposes, whereas the conditional formulation concerns how much evidence scientists should have before deciding to use T for scientific purposes. The attributive formulation is reticent about the epistemic conditions under which scientists can accept T, whereas the conditional formulation is reticent about what scientists believe when they decide to use T for scientific purposes. In this respect, the conditional and attributive formulations complement each other.

How do conditional realism and empiricism differ from epistemic realism and empiricism? Epistemic realism and empiricism are rival views about how much evidence we have for T. Epistemic realism asserts that we have sufficient evidence for believing that T is true, whereas epistemic empiricism says that we have sufficient evidence for believing that T is EA. By contrast, conditional realism and empiricism are rival views about how much evidence scientists should have to accept T, i.e., they are rival views about under what epistemic conditions scientists can decide to use T for scientific purposes. Conditional realism asserts that scientists should have evidence for believing that T is true, while conditional empiricism says that they should have evidence for believing that T is EA. Both conditional realism and empiricism are neutral about whether scientists actually have evidence for the belief that T is true or for the belief that T is EA.

Epistemic realists and empiricists are currently engaged in an active debate about how much T is supported by evidence. For example, epistemic empiricists pose the problem of underdetermination to show that we do not have evidence for believing that T is true, although we have evidence for believing that T is EA. Epistemic realists counter that theoretical virtues, such as simplicity and wide scope, are epistemic virtues, so we have evidence for the truth of T. Once the debate between epistemic realists and empiricists ends, we may be able to tell whether we have evidence for believing that T is true or only for believing that T is EA.

The epistemic, attributive, and conditional formulations of rival positions attempt to reveal different aspects of science. The epistemic formulation attempts to reveal how much evidence we have for T. The attributive formulation attempts to reveal what scientists, viz., those who commit themselves to confront the world in terms of T, believe regarding T. The conditional formulation attempts to reveal how much evidence scientists should have for T before deciding to use it for scientific purposes.

Let me bring out an implication of the fact that these three kinds of formulations concern different aspects of science, and thus they generate different kinds of debates. You should explore a debate depending on which aspect of science you are interested in. If you are interested in whether there is sufficient evidence for T, you should choose the debate between epistemic realists and empiricists. If you are interested in what accepters of T believe regarding T, you should study the debate between attributive realists and empiricists. If you are interested in how much evidence scientists need to have for T before deciding to use it for scientific purposes, you should explore the debate between conditional realists and empiricists. It is wrong to explore, for example, the debate between attributive realists and empiricists when you are interested in whether we have sufficient evidence for T.

Now, contemporary science abounds with theories. Do we have sufficient evidence for thinking that, say, the oxygen theory is true or EA? If so, what is it? Do we have sufficient evidence for believing that, say, string theory is true or EA? If not, why not? What makes us think that we have sufficient evidence for the belief that the oxygen theory is true or EA, but not for the belief that string theory is true or EA? How can we go about determining whether particular theories in current science are true or EA? In other words, how do we distinguish true theories from false ones and distinguish EA theories from those that are not?

Rival answers to these questions can be found in epistemic realism and empiricism, which state, respectively, that *successful* theories are true and EA. Larry Laudan gives a clear definition of success: "a theory is successful if it makes substantially correct predictions, if it leads to efficacious interventions in the natural order, if it passes a battery of standard tests" (Laudan, 1981, p. 23). The oxygen theory is successful in this sense. By contrast, string theory is not successful in that it has yet to be confirmed. For now, it is merely a set of assumptions. Thus, epistemic realists and empiricists would say that we have evidence for believing that the oxygen theory is true and EA, respectively, but not for believing that string theory is true and EA. In short, we can pick out true and EA theories from current science, depending on whether they are successful (Park, 2020a, p. 488).

No answers to the questions above can be found in either attributive realism and empiricism, which assert that scientists believe that T is true and EA, respectively. These two positions are irrelevant to the issue of which theories from contemporary science are true or EA. They are instead relevant to the issue of whether the commitment to using theories for scientific purposes involves the belief that they are true or EA. Thus, they are silent about whether a particular current theory is true or EA. It is, therefore, natural that attributive realism is classified as a variant of putative realism in Sect. 5.5 of Chap. 5.

No answers to the questions above can be found in conditional realism and empiricism, which assert that scientists should only accept T, provided that they first have evidence for the beliefs that it is true and EA, respectively. These two positions are irrelevant to the issue of which theories in current science are true and EA. They instead concern the issue of what epistemic requirements scientists should meet to accept T. They claim that scientists should have evidence for the truth and the empirical adequacy of T, respectively. Consequently, they imply that scientists can accept string theory if they have evidence for the beliefs that T is true and EA, respectively. However, they are silent about whether scientists can accept T, i.e., how much evidence they actually have for it. It follows that conditional realism is also a variant of putative realism.

Now, suppose that you are interested in what the world looks like, and you are led to wonder whether you can trust what science says about the world. For example, you are interested in whether the universe came into existence about 13.8 billion years ago and whether there are 11 dimensions. This being the case, you should explore the debate between epistemic realists and empiricists. You should not explore the debate between attributive realists and empiricists or the debate between conditional realists and empiricists; you would only waste your time if you were to explore either of these two debates.

The name "realism" is attached to epistemic realism, attributive realism, and conditional realism. However, they are radically different theses about science, and thus they should be strictly differentiated. Epistemic realists claim that T is warranted, whereas attributive realists and conditional realists may claim that T is unwarranted. After all, it is coherent for one to claim that T is unwarranted while claiming that scientists believe that T is true; such an agent is an epistemic antirealist but attributive realist. It is also coherent for one to claim that T is unwarranted while claiming that scientists can only accept T if they first have evidence for the belief that it is true. Such an agent is an epistemic antirealist and conditional realist. We should not be misled by the word "realism." The same point applies mutatis mutandis to epistemic empiricism, attributive empiricism, and conditional empiricism.

## 6.5 A Pragmatic Advantage of the Epistemic Formulation

The epistemic formulation has a pragmatic advantage over the attributive and conditional formulations: indispensabilists in philosophy of mathematics can rely on the epistemic formulation, but not on the other formulations. Indispensabilists believe that the mathematical constituents of our best theories are true, arguing that they are confirmed, like nonmathematical constituents, by observational evidence (Quine, 1948; Putnam, 1971). Note that they do not believe that the mathematical elements of *any* scientific theory are true; instead, they believe that the mathematical elements of our *best* theories are true. But which theories from contemporary science are the best ones? Is the oxygen theory one of them? What about string theory? The answers to these questions can be derived from the epistemic formulation, but not from the attributive and conditional formulations.

According to epistemic realism, the success of T is sufficient evidence for the truth of T. On this account, as long as T is successful, it is one of our best theories. The oxygen theory is successful, while string theory is not. As a result, epistemic realists would say that the oxygen theory is one of our best theories, whereas string theory is not. Indispensabilists, appealing to epistemic realism, would believe that the mathematical constituents of the oxygen theory are true, but would not believe that the mathematical components of string theory are true. Epistemic realism is, therefore, useful for indispensabilists (Park, 2020a, p. 488).

According to attributive realism, accepters of T believe that it is true. Can we accept the oxygen theory and string theory? Attributive realism has no answer to this question, given that it does not make any claim about the epistemic condition under which we can accept T, to say nothing of whether any current theory has met that epistemic condition. Again, attributive realism is not an epistemic thesis but rather a descriptive thesis regarding what accepters of T believe. Hence, it makes no assertion about whether we can justifiably believe the oxygen theory or string theory. For this reason, it is not useful to indispensabilists (Park, 2020a, p. 488).

What about conditional realism? It asserts that scientists can accept T, provided that they have evidence for its truth. But do scientists have evidence for the truth of the oxygen theory or string theory? Conditional realism does not have an answer to this question. It is not a thesis about how much evidence we have for T, and hence it is not a thesis about which theories in current science are believable. Rather, it is a thesis about the epistemic condition under which we can accept T. It makes no assertion about whether any current theory has met the epistemic condition. Therefore, it is not useful to indispensabilists.

## 6.6 Conceptual and Empirical Objections

This section exposes conceptual and empirical problems with conditional realism and empiricism. Let me begin with a conceptual one.

The conceptual problem can be exposed with the use of the following analogy. Imagine that an admissions officer of a university says that high school students should only apply for her university, provided that they have already been admitted into it. Obviously, this contention puts the cart before the horse. Applying to a university is a prior condition of being admitted to it, not the other way around. Consequently, the admissions officer should rather say that high school students can be admitted to her university, provided that they first apply to it.

Conditional realists and empiricists make a similar error. Scientists' adoption of the policy of using T for scientific purposes is a prior condition for their obtaining evidence for T, not the other way around. Consequently, it is conceptually problematic to suggest that scientists should only accept T, provided that they first have evidence for the beliefs that T is true or EA. By contrast, it is conceptually sound to suggest that scientists can obtain evidence for the beliefs that T is true and EA, provided that they first accept T, i.e., provided that they first decide to use T for scientific purposes.

Under what conditions would conditional realism and empiricism make sense? Imagine that scientists do not have evidence for T at all. God gives a group of scientists evidence for believing that T is true, while giving another group of scientists evidence for believing that T is EA. Under these circumstances, conditional realists would say that the first group of scientists, but not the second, can decide to use T for scientific purposes. By contrast, conditional empiricists would say that both groups of scientists can decide to use T for scientific purposes. However, it is implausible that God gives scientists the evidence. I leave to conditional realists and empiricists the task of showing under what condition their proposals make sense.

Let me now raise an empirical objection to conditional realism and empiricism. As Laudan (1981, p. 33) observes, many theories have been discarded during the history of science. They may have appeared to be true and EA, but they turned out to be false and not EA (Park, 2001, p. 78; Lyons, 2003, p. 898). We can now see that past scientists did not have evidence for the beliefs that their theories were true and EA. They rather had evidence, at best, for the truth of some theoretical assumptions or for the limited use of their theories. However, they accepted their theories and used them for scientific purposes. Conditional realism and empiricism imply, however, that they should not have accepted their theories. Hence, conditional realism and empiricism clash with the history of science. If past scientists had adopted conditional realism and empiricism as their philosophical frameworks, they would have made no scientific progress.

Conditional empiricists may argue that conditional empiricism imposes a lighter burden on scientists than does conditional realism, so conditional empiricism is better than conditional realism. In my view, however, both conditional realism and empiricism impose such a heavy burden on scientists that there is not much to gain by saying that conditional empiricism imposes a lighter burden on scientists than conditional realism. Both conditional realism and empiricism are philosophical doctrines that would only stifle the growth of science.

## 6.7 Doxastic Conditionalism

The history of science motivates me to consider an alternative to conditional realism and empiricism, which I call "doxastic instrumentalism." It holds that scientists can decide to apply T to the world, provided that they *believe* that it will be *useful*. Imagine that a scientist has conceived a new scientific theory. It has not yet been tested, but she evaluates it by using her background knowledge, and she believes that it will be useful. Under this circumstance, she can decide to apply it to the world. Suppose, by contrast, that she believes that it will be useless after evaluating it in light of her background knowledge. Under this circumstance, she should decide not to apply it to the world.

Doxastic instrumentalism imposes a far lighter requirement on scientists than do conditional realism and empiricism. It does not even require that T should be useful. It only requires that scientists should *believe* that T will be useful. Only if they believe that it will be useful can they decide to apply it to the world. Because doxastic instrumentalism imposes such a light requirement, it is compatible with the history of science, during which false and empirically inadequate theories were applied to the world. Although the theories were false and not EA, past scientists could have believed that they would be useful. The refutation of doxastic instrumentalism requires not the thesis that past theories were useless but rather the thesis that past scientists did not believe that they would be useful.

Is doxastic instrumentalism tenable? On close examination, it is not. As we noted in Sect. 3.5 of Chap. 3, it is subject to three problems: Moore's paradox, the problem of disconcerting questions, and the problem of deceptive speech acts. These three problems arise, you may recall, when we apply a theory to the world without believing it to be true. They indicate that we should believe that T is true before we decide to use it for scientific purposes. However, doxastic instrumentalism suggests that we should believe that T will be *useful* before deciding to apply it to the world. Therefore, we can conclude that doxastic instrumentalism imposes too light a doxastic requirement on scientists.

This consideration motivates me to consider an alternative to doxastic instrumentalism that I call "doxastic realism." It holds that scientists can decide to apply T to the world, provided that they first believe that it is *true*. Imagine again that a scientist conceives of a new scientific theory. It has not yet been tested, but she evaluates it using her background knowledge, and she believes that it is true. Under this condition, she can decide to apply it to the world. Suppose, by contrast, that she believes that it is false after evaluating it in light of her background knowledge. Under this condition, she cannot decide to apply it to the world.

However, doxastic realism is untenable. In Sect. 3.6 of Chap. 3, we noted that antirealists can use T to predict observational events, even when they do not believe it, provided that they know that it has made true predictions. They only need to believe that T will continue to make true predictions. What if T was never used to make a prediction before? Only realists can use it to make a prediction. If you make a prediction, despite not believing T, you cannot persuade your epistemic colleagues that the prediction is true, and they will not spend their time and energy in ascertaining the prediction. In any event, antirealists need only believe that T will make true predictions, contra doxastic realism, provided that they know that it has made true predictions.

The preceding discussion of doxastic realism and instrumentalism leads us to a new proposal, which I call "doxastic conditionalism." According to doxastic conditionalism, scientists can have different beliefs about T, depending on whether they plan to use T to explain or predict phenomena and depending on whether they know that T has already made true predictions. They should believe that T is true before deciding to explain events, whether or not they know that it has already made true predictions. They should also believe that it is true before deciding to predict events, in case they do not know that T has already made true predictions. By contrast, they do not need to believe that T is true, and they can instead believe that it will continue to make true predictions, in case they know that it has already made true predictions and they plan to use it to make predictions.

Doxastic conditionalists comply with the epistemic imperative: "Act only on an epistemic maxim through which you can at the same time will that it should become a universal one" (Park, 2018, p. 441). Imagine that doxastic conditionalists have a positive philosophical theory, whatever it is, and that they decide to use it in their philosophical endeavors. They believe that it is true when deciding to use it to give explanations. By contrast, they choose between the belief that it is true and the belief that it will continue to make true predictions, depending on whether they know that it has made true predictions. In short, doxastic conditionalism imposes such a reasonable requirement on scientists that they happily impose it on themselves.

By contrast, imagine that conditional realists have a philosophical theory, whatever it is, and that they use it in their philosophical endeavors. Scientists come along and argue that conditional realists should have evidence for the truth of their philosophical theory before deciding to use it in their philosophical endeavors. Conditional realists would protest that scientists impose a heavy requirement on them, pointing out that they cannot obtain evidence for the truth of their philosophical theory without first using it. In this case, conditional realists would violate the epistemic imperative. They believe that scientists should act on the maxim, "Adopt the policy of using T for scientific purposes, provided that you have evidence for the truth of it," despite the fact that they do not will the maxim to become a universal one.

A similar objection can be raised against conditional empiricists. Imagine that they propose that scientists should act on the maxim, "Adopt the policy of using T, provided that you have evidence for believing that T is EA." Scientists come along and tell conditional empiricists to act on the maxim themselves. Conditional empiricists refuse to act on the maxim, complaining that scientists impose a heavy requirement on them. In such a case, they violate the epistemic imperative.

How does doxastic conditionalism relate to epistemic realism, the view that we have sufficient evidence for the belief of T? Strictly speaking, they are compatible with each other. Epistemic realists can maintain without contradicting themselves that scientists who apply T to the world can choose between the belief that T is true and the belief that it will continue to make true predictions, depending on whether scientists know that it has made true predictions. However, epistemic realists would say that scientists are overly cautious about T if scientists choose the belief that T will continue to make true predictions over the belief that it is true.

How does doxastic conditionalism relate to attributive realism and empiricism? It is worth noting the following three differences between them. The first difference is that attributive realism and empiricism are descriptive theses about what accepters of T believe, whereas doxastic conditionalism is a normative thesis about what scientists should believe before deciding to apply T to the world.

The second difference is that doxastic conditionalism is sensitive, while attributive realism and empiricism are insensitive, to what accepters of T plan to do with T. Accepters of T might plan to explain or predict phenomena with it. Doxastic conditionalism distinguishes between the two different plans, asserting that accepters of T might have a choice between the belief that T is true and the belief that T will continue to make true predictions, in case they plan to predict phenomena. By contrast, attributive realism and empiricism do not distinguish between the two different plans, asserting that accepters of T believe that T is true and EA, respectively, regardless of whether their plan is to explain or predict events. Hence, attributive realism and empiricism overlook the fact that accepters of T might change their beliefs, depending on whether they plan to use T to explain or predict events.

The third difference is that the word "acceptance" figures prominently in the formulation of attributive realism and empiricism, whereas it does not in the formulation of doxastic conditionalism. I deliberately avoided using the word "accept" and used instead the words "decide" and "plan" in the formulation of doxastic conditionalism. The reason for avoiding "accept" will become clear in the following section.

## 6.8 Acceptance

Van Fraassen and Dellsén formulate realism and empiricism in terms of "acceptance." This section argues against the use of this word.

The word "accept" can be eliminated from Dellsén's formulation without distorting the idea behind it. An alternative formulation states that conditional realism and empiricism are positions that scientists should only *adopt* the policy of using T for scientific purposes, provided that they first have evidence for the beliefs that T is true and EA, respectively. Note that the word "accept" is replaced with the phrase "adopt the policy of using T for scientific purposes." Such alternative formulations are clearer than Dellsén's formulation, in which the word "accept" figures.

In Sect. 6.5 of this chapter, I claimed that Dellsén's formulation has the conceptual problem of putting the cart before the horse. In my view, Dellsén would not have made such a mistake if he had used the alternative phrase "adopt the policy of using T for scientific purposes" instead of the word "accept." After all, the conceptual problem is evident in the sentence, "Scientists should only adopt the policy of using T for scientific purposes, provided that they first have evidence for the beliefs that T is true and EA," but not in the sentence, "Scientists should only accept T, provided that they have evidence for the beliefs that T is true and EA" due to the obscure word "accept." This obscure word has the effect of concealing the conceptual problem.

A similar criticism can be directed at van Fraassen's formulation. The word "accept" can be purged from his formulation without distorting the idea behind it. An alternative formulation of attributive realism and empiricism is that according to attributive realism, *the commitment* to using T for scientific purposes involves the belief that T is true and that according to attributive empiricism, *the commitment* to using T for scientific purposes involves the belief that T is EA. Note that the original word "accept" has been replaced with the alternative phrase "the commitment to using T for scientific purposes." In my view, this alternative formulation is clearer.

In addition, debates would have seen fewer mistakes in interpreting attributive empiricism, had van Fraassen advanced the alternative formulation above. For example, Papineau interprets attributive empiricism as promoting the epistemic thesis that "we ought never to believe in the truth of any theory which goes beyond the observable phenomena" (Papineau, 1996, p. 8). Van Fraassen retorts that "I do not advocate agnosticism about the unobservable" (van Fraassen, 2007, p. 343; 2017, p. 99). Papineau would not have misunderstood attributive empiricism if van Fraassen had formulated it as the view that "the commitment to using T for scientific purposes involves the belief that it is EA" instead of as the view that "acceptance of a theory involves as belief only that it is empirically adequate" (van Fraassen, 1980, p. 12). The alternative formulation makes explicit that attributive empiricism is not an epistemic thesis but rather a descriptive thesis. As a result, Papineau would not have superimposed the epistemic thesis on attributive empiricism.

One caveat is in order. I do not deny that the ideas behind the formulations invoking "acceptance" are valuable. As I argued in Sect. 6.4 of this chapter, they reveal aspects of science that the epistemic formulation does not and therefore help us learn more about science. I only claim that van Fraassen's and Dellsén's formulations can dispense with "acceptance" and that the formulations without this obscure word are less misleading.

## 6.9 Summary

There are formulational and epistemological debates concerning realism. Putnam's (1975) formulation has promoted both kinds of debates. By contrast, van Fraassen's (1980) formulation has promoted formulational debates, but not epistemological ones. In addition, there are wide and narrow formulations. Putnam's (1975) formulation is wide, whereas van Fraassen's (2017) formulation is narrow. Consequently, there can be more debates between rival discussants under Putnam's formulation than under van Fraassen's formulation.

The epistemic, attributive, and conditional formulations attempt to reveal different aspects of science. If you are interested in what epistemic attitude you should take toward T, you should explore the debate between epistemic realists and empiricists. If you are interested in what scientists believe regarding T when they commit themselves to use it for scientific purposes, you should explore the debate between attributive realists and empiricists. If you are interested in how much evidence scientists should have to adopt the policy of using it for scientific purposes, you should explore the dispute between conditional realists and empiricists.

Scientists can obtain evidence for the beliefs that T is true or EA after accepting T, not the other way around. In addition, the history of science suggests that scientists did not have evidence for the beliefs that their best theories were true and EA. For these reasons, it is wrong to propose that scientists should only accept T, provided that they first have evidence for the beliefs that T is true or EA, and it is right to say that scientists can have evidence for T, provided that they first accept it.

My alternative proposal is as follows. If scientists plan to use T to *explain* events, they should believe that T is true, whether or not they know that it has been already used to make true predictions. If they plan to use T to *predict* events, they may believe that T is true, or that T will continue to make true predictions, depending on whether they know that T has already made true predictions. If they have the knowledge, they may only believe that T will continue to make true predictions. If they do not have the knowledge, however, they should believe that T is true. Finally, those who plan to formulate realism and empiricism in terms of "acceptance" are asked to explain what the benefits are of doing so.

## References

Dellsén, F. (2017). Reconstructed empiricism. *Acta Analytica, 32*(1), 95–113.
Healey, R. (2019a). Four points in response to Seungbae Park. *Social Epistemology Review and Reply Collective, 8*(12), 43–45.
Healey, R. (2019b). The aims of reliable knowledge: Reply to Seungbae Park. *Social Epistemology Review and Reply Collective, 8*(9), 25–30.
Laudan, L. (1981). A confutation of convergent realism. *Philosophy of Science, 48*(1), 19–49.
Leplin, J. (1997). *A novel defense of scientific realism*. Oxford University Press.
Lyons, T. (2003). Explaining the success of a scientific theory. *Philosophy of Science, 70*(5), 891–901.
Lyons, T. (2017). Epistemic selectivity, historical threats, and the non-epistemic tenets of scientific realism. *Synthese, 194*(9), 3203–3219.
Musgrave, A. (1985). Realism vs. constructive empiricism. In P. Churchland & C. Hooker (Eds.), *Images of science: Essays on realism and empiricism* (pp. 197–221). University of Chicago Press.
Musgrave, A. (2017). Strict empiricism versus explanation in science. In E. Agazzi (Ed.), *Varieties of scientific realism: Objectivity and truth in science* (pp. 71–93). Springer.
Nickles, T. (2017). Cognitive illusions and nonrealism: Objections and replies. In E. Agazzi (Ed.), *Varieties of scientific realism: Objectivity and truth in science* (pp. 151–163). Springer.
Papineau, D. (1996). *The philosophy of science*. Oxford University Press.
Park, S. (2001). *Scientific realism vs. scientific antirealism*. Ph.D. dissertation. University of Arizona.
Park, S. (2011). Coherence of our best scientific theories. *Foundations of Science, 16*(1), 21–30.
Park, S. (2018). In defense of the epistemic imperative. *Axiomathes, 28*(4), 435–446.
Park, S. (2019a). Constructive empiricism in a social world: Reply to Richard Healey. *Social Epistemology Review and Reply Collective, 8*(10), 146–154.
Park, S. (2019b). How to formulate scientific realism and antirealism. *Journal for General Philosophy of Science, 50*(4), 477–488.
Park, S. (2020a). Formulational vs. epistemological debates concerning scientific realism. *Dialogue: Canadian Philosophical Review, 59*(3), 479–496.
Park, S. (2020b). Replies to Healey's comments regarding van Fraassen's positions. *Social Epistemology Review and Reply Collective, 9*(1), 38–47.
Park, S. (2020c). How to overcome antirealist objections to scientific realism. *Axiomathes, 30*(1), 1–12.
Putnam, H. (1971). *Philosophy of logic*. Harper and Row.
Putnam, H. (1975). *Mathematics, matter and method: Philosophical papers, volume 1*. Cambridge University Press.
Quine, W. (1948). On what there is. *Review of Metaphysics, 2*(5), 21–38.
van Fraassen, B. (1980). *The scientific image*. Oxford University Press.
van Fraassen, B. (1989). *Laws and symmetry*. Oxford University Press.
van Fraassen, B. (2007). Reply: From a view of science to a new empiricism. In B. Monton (Ed.), *Images of empiricism: Essays on science and stances, with a reply from Bas van Fraassen* (pp. 337–383). Oxford University Press.
van Fraassen, B. (2017). Misdirection and misconception in the scientific realism debates. In E. Agazzi (Ed.), *Varieties of scientific realism: Objectivity and truth in science* (pp. 95–108). Springer.
Vickers, P. (2017). Understanding the selective realist defence against the PMI. *Synthese, 194*(9), 3221–3232.
Wray, B. (2008). The argument from underconsideration as grounds for anti-realism: A defence. *International Studies in the Philosophy of Science, 22*(3), 317–326.
Wray, B. (2013). The pessimistic induction and the exponential growth of science reassessed. *Synthese, 190*(18), 4321–4330.

# Chapter 7
# Scientific Realism and Scientific Understanding

**Abstract** According to epistemism, we scientifically understand explananda in terms of explanantia, provided that they are true and we justifiably believe them. On this account, scientific understanding requires the three ingredients of knowledge: belief, justification, and truth. Therefore, scientific understanding is attainable for realists, but not for antirealists. According to anti-epistemism, scientific understanding requires explanation and prediction, but none of the three ingredients of knowledge. I object that anti-epistemists have the burden of giving an account of explanation and prediction without appealing to the three ingredients of knowledge and an account of when misunderstanding arises.

**Keywords** Acceptance · Epistemism · Belief · Empiricism · Justification · Truth · Models · False theories · Understanding

## 7.1 Introduction

The nature of scientific understanding is an emerging topic for epistemologists and philosophers of science. This chapter explores how realism relates to scientific understanding. I defend the position that I call "epistemism." It holds that scientific understanding requires belief, justification, and truth. Given that realists affirm, while antirealists deny, that T is warranted, epistemism implies that scientific understanding is attainable for realists, but not for antirealists. This disadvantage of antirealism can be added to the pool of disadvantages of antirealism that I unfolded in Chap. 3.

Epistemism would be embraced by Stephen Grimm (2006, 2014), who argues that understanding is "a species of knowledge" (Grimm, 2006, p. 516). However, it would be resisted by Finnur Dellsén (2016, 2017a), Catherine Elgin (2004, 2012), and Henk de Regt (2015). Dellsén argues that neither belief nor justification is required for scientific understanding. Elgin claims that models, though false, nonetheless generate scientific understanding. De Regt maintains that past and present


S. Park, *Embracing Scientific Realism*, Synthese Library 445,
https://doi.org/10.1007/978-3-030-87813-9_7

theories, though false, nevertheless give rise to scientific understanding. These three philosophers' arguments are critically examined in this chapter.

This chapter is structured as follows. In Sect. 7.2, I criticize Dellsén's (2016, 2017a) suggestion that belief is not required for understanding. In Sect. 7.3, I raise difficulties against Dellsén's (2016, 2017a) view that justification is not required for understanding. In Sect. 7.4, I critically examine Elgin's (2012) and de Regt's (2015) assertion that truth is not required for understanding. The central thesis of this chapter is that realists understand the world more than antirealists.

## 7.2 No Belief?

### *7.2.1 Belief and Understanding*

There is a connection between explanation and understanding. We come to understand why an event occurs once we are provided with an explanation of why it occurs. For example, we come to understand why a lunar eclipse occurs once we are told that it happens due to the alignment of the three objects: the sun, the earth, and the moon. As we saw in Sect. 5 of Chap. 3, however, antirealists cannot explain events in terms of T because they do not believe T. If they explain events in terms of T, they run into one of the three problems: Moore's paradox, the problem of disconcerting questions, and the problem of deceptive speech acts. Therefore, antirealists cannot help their epistemic colleagues understand why an event occurs in terms of T.

Can antirealists themselves understand why an event occurs in terms of T? My answer is "No." Suppose that some antirealists have never been exposed to contemporary astronomy,[1] so they do not understand why a lunar eclipse occurs. Realists say to them, "A lunar eclipse occurs because three objects—the sun, the earth, and the moon—are aligned." This explanation cannot give rise to understanding in antirealists' minds because they do not believe the explanation. If antirealists acquire the understanding, that is in part because they believe the explanation. Once they believe the explanation, however, they cease to be antirealists.

What if antirealists insist that they understand why a lunar eclipse occurs even if they do not believe realists' explanation? I would encourage them to consider the following imaginary dialogue between a realist and an antirealist:

A: I don't understand why a lunar eclipse occurs.
R: Let me explain. It occurs due to the alignment of three objects: the sun, the earth, and the moon.
A: I now understand why it occurs, but I don't believe that it occurs due to the alignment of the three objects.

[1] In this chapter, I assume that antirealists do not believe what contemporary astronomy tells us about the mechanism that brings about a lunar eclipse.

R: What you just said is puzzling. How can you say, "I now understand why it occurs, but I don't believe that it occurs due to the alignment of the three objects"?

Why is it peculiar for the antirealist to say, "I now understand why it occurs, but I don't believe that it occurs due to the alignment of the three objects"? The sentence, "I now understand why it occurs" indicates that the antirealist now believes that a lunar eclipse occurs due to the alignment of the three objects (Park, 2016, pp. 78–79). As a result, the realist assumes that the antirealist believes that a lunar eclipse occurs due to the alignment of the three objects. Contrary to the realist's assumption, however, the antirealist declares that she does not believe that a lunar eclipse occurs due to the alignment of the three objects. In short, the realist's assumption does not match up with what the antirealist says. As a result, the realist is puzzled over what the antirealist says.

What can we learn from this imaginary dialogue between the realist and the antirealist? Belief is an essential constituent of understanding. The antirealist cannot understand why a lunar eclipse occurs because she does not believe that a lunar eclipse occurs due to the alignment of the three objects. This claim about the requirement of understanding accords with epistemism, according to which understanding requires belief, justification, and truth. However, it clashes with Dellsén's account of understanding, to which I turn now.

### 7.2.2 Acceptance and Understanding

Dellsén contends that belief is not an ingredient of understanding. We can understand an explanandum in terms of explanantia, although we do not believe any of the explanandum and explanantia. We can instead *accept* them. To understand this cryptic claim, we need to be clear about Dellsén's technical notion of acceptance.

For Dellsén, belief and acceptance are significantly different from each other. He (2017a, pp. 248–249) relies on Jonathan Cohen's (1992) account of the difference between belief and acceptance, according to which to believe a proposition is to feel it to be true, whereas to accept it is to include it "among one's premises for deciding what to do or think in a particular context" (Cohen, 1992, p. 4). Cohen uses the example of a lawyer to explicate the difference between belief and acceptance. Suppose that you are a lawyer and that your client tells you the truth that she committed a crime. As a result, you feel that she is guilty, i.e., you *believe* that she is guilty. In court, however, you *accept* that your client is not guilty and then speak to the jurors as if your client was not guilty. You use as a premise the proposition that your client is not guilty whenever you make an inference and speak in court.

For Dellsén, to accept a scientific theory is to adopt the policy of using "the theory for various scientific purposes, e.g., as the basis for empirical predictions, as premises in one's explanations, and as a guide in further theory construction" (Dellsén, 2017b, p. 99). Note that according to him, we can use T for scientific purposes after merely accepting it, which implies that we can explain and predict events in terms of T, even if we do not believe that T is true, i.e., even if we do not feel that T is true. In this spirit, he suggests that "understanding something may merely involve treating certain propositions or theories as given in the context of explaining something, as opposed to being disposed to feel that the propositions or theories are true" (Dellsén, 2017a, p. 248). In short, we can understand events in terms of T although we do not believe that T is true.

Bas van Fraassen (1980) would concur with Dellsén on this count. For van Fraassen, to accept T is to commit oneself to confront the world in terms of T. To confront the world in terms of T includes understanding the world in terms of T. For example, geologists confront geological phenomena in terms of plate tectonics. When they do so, they understand the existence of sea life fossils on Mt. Everest in terms of plate tectonics. For constructive empiricists, the acceptance of plate tectonics involves the belief that it is empirically adequate (EA), but not that it is true. Accordingly, van Fraassen would say that geologists believe that plate tectonics is EA, but not that it is true. This claim about what geologists believe implies that geologists understand geological phenomena in terms of plate tectonics, although they do not believe that it is true. This implication of van Fraassen's notion of acceptance is consistent with Dellsén's view that belief is not required for understanding.

Let me argue against Dellsén's view that understanding requires not belief but acceptance. Imagine that you travel in a time machine to ancient China. You see the ancient Chinese shoot fireworks into the sky during a lunar eclipse. You ask them why they shoot fireworks into the sky. They answer that they are trying to get the moon back from a dragon that ate it. The ancient Chinese just provided you with their hypothesis explaining why a lunar eclipse occurs. As a result, you acquire an understanding of why they behave as they do. However, you do not develop a new understanding of why a lunar eclipse occurs. You were previously armed with the modern explanation that a lunar eclipse occurs due to the alignment of three objects: the sun, the earth, and the moon. The modern explanation makes it psychologically impossible for you to understand a lunar eclipse in terms of the dragon hypothesis.

Of course, you can *accept* the dragon hypothesis, and as a result, you can speak and behave as if it were true that a dragon swallowed the moon. For example, you can say to ancient Chinese, "I agree that the moon disappeared because a dragon devoured it," and then shoot off fireworks into the sky just as they do to form a social bond with them. You can *pretend* that you understand a lunar eclipse in terms of their hypothesis. However, you cannot understand a lunar eclipse yourself in terms of the ancient Chinese hypothesis. In the back of your mind, you think that they have misunderstood why a lunar eclipse occurs and that it is irrational to shoot fireworks into the sky to free the moon from a dragon. The same is true of the example of the lawyer. She can pretend that her client is not guilty; in the back of her mind, however, she thinks that her client is guilty. Consequently, we can conclude that when there is no belief, there is no understanding, and there is instead pretending.

Let me raise another difficulty against Dellsén's suggestion that we can understand events in terms of T only by accepting it. Under what condition does misunderstanding arise? You do not have an answer to this question under Dellsén's framework that acceptance rather than belief is crucial to understanding (Park, 2017a, p. 386). You cannot say that misunderstanding arises if and only if acceptance is false, for acceptance is neither true nor false but can only be either useful or useless. For example, if the prosecutor did not have proof that the lawyer's client is guilty, the lawyer's acceptance of the proposition that her client is not guilty would be useful. By contrast, if the prosecutor had proof, the lawyer's acceptance would be useless. Similarly, if your acceptance of the ancient Chinese hypothesis helped you have a good relationship with ancient Chinese, it would be useful. By contrast, if it did not help you at all, it would be useless. In short, it is wrong to suggest that misunderstanding occurs when acceptance is false.

Epistemism has a natural answer to the question: under what condition does misunderstanding arise? The answer is that it arises when a belief is false. For example, the ancient Chinese misunderstood a lunar eclipse because they had the false belief that a dragon swallowed the moon. The fundamental difference between understanding and misunderstanding is that understanding corresponds to the world, whereas misunderstanding does not (Park, 2017a, p. 386). A mental state that corresponds to the world is a true belief, and a mental state that does not is a false belief. Therefore, it is natural to suggest that a true belief and a false belief are essential components of understanding and misunderstanding, respectively.

My thesis that belief is indispensable to scientific understanding can be strengthened by considering the following dialogue between two children:

> Child 1: Who wins if Superman and Batman fight with each other?
> Child 2: Of course, Superman wins. After all, he can lift up a heavy rock as large as a mountain. Moreover, he can fly. But Batman can do none of these things.
> Child 1: No, Batman wins. He only needs to carry Kryptonite.
> Child 2: I understand. (Park, 2017a, p. 384)

Note that the second child comes to *understand* how Batman can beat Superman as a result of the first child's explanation. However, this kind of understanding is not scientific but fictional (Park, 2020, p. 178). A fundamental difference between scientific and fictional understanding is that beliefs are involved in scientific understanding, but not in fictional understanding. The two children do not believe the propositions about Superman and Batman. They merely accept them to have some fun. To say that acceptance rather than belief is essential to scientific understanding is to demote scientific understanding to the level of fictional understanding.

Let me now turn to an ethical issue. There are cases in which it is unethical merely to accept a proposition and then to act on it. Imagine that you are on trial in court. The judge does not believe that you are guilty due to the lack of evidence, but she accepts that you committed a crime. She declares, "I hereby sentence you to life in prison." You protest that it is immoral for her to issue the sentence. She replies that she accepted the proposition that you are guilty for the purpose of issuing the sentence. Such an example indicates that there are cases in which it is immoral to accept a proposition and then to act on it, although there might be cases in which it is moral to accept a proposition and then to act on it. Therefore, theorists, such as Dellsén and van Fraassen, owe us an account of under what condition it is permissible and under what condition it is impermissible to accept a proposition and then to act on it (Park, 2020, pp. 182–183).

Dellsén and van Fraassen might advance the following proposal. The judge should not accept the proposition that you committed a crime because accepting the proposition makes you unhappy. However, we may accept a scientific theory because accepting it does not make anyone unhappy. For example, I may accept plate tectonics and then say to my children, although I do not believe it, that Mt. Everest is rising because the Indian plate and the Eurasian plate are pushing up against each other. Accepting a scientific theory does not harm anyone. So might Dellsén and van Fraassen suggest.

However, it is problematic to suggest that accepting a scientific theory does not harm anyone. As we have seen in Sect. 3.5 of Chap. 3, when you speak what you do not believe, you are disregarding others' desire to acquire truths and avoid errors. Moreover, imagine that scientists accept the theory of cold fusion, a false theory, according to which nuclear reactions can occur at room temperature. They do not believe it, but they talk as if it were true in their research proposal so that they can receive research funds. They receive the funding only to waste it.

So far, I have criticized Dellsén's view that acceptance, rather than belief, is a component of scientific understanding. My criticism has a negative impact on his account of scientific progress. He proposes that "scientific progress consists in increasing understanding" (Dellsén, 2016, p. 73). If understanding were devoid of belief, however, scientific understanding would be similar to fictional understanding. Consequently, understanding is not an adequate explanans to account for progress in science, although it might be an adequate explanans to account for progress in science fiction (Park, 2020, p. 178).

### 7.2.3 Empiricism and Understanding

In Sect. 1.9 of Chap. 1, I defined as empiricism the position that T is EA. Empiricists might put forward the following argument. Explanation and understanding require not belief in truth but rather belief in empirical adequacy. Empiricists believe that T is EA. As a result, they can explain events in terms of T, they can help their epistemic colleagues understand events in terms of T, and they themselves can understand events in terms of T. This subsection aims to refute the premise of this possible argument that explanation and understanding require not belief in truth but rather belief in empirical adequacy.

Van Fraassen thinks that empiricists can explain events in terms of T. He writes, "acceptance of a theory involves as belief only that it is empirically adequate" (van Fraassen, 1980, p. 12), and if "the acceptance is at all strong, it is exhibited in the person's assumption of the role of explainer, in his willingness to answer questions *ex cathedra*" (van Fraassen, 1980, p. 12). This view about the role of explainer can be illustrated by the example of the ancient Chinese hypothesis that a dragon swallows the moon. Once empiricists accept this hypothesis, they are willing to use it to answer the question, "Why does a lunar eclipse occur?" Specifically, they are willing to say, "A lunar eclipse occurs because a dragon swallows the moon." When they give this explanation, they believe that the dragon hypothesis is EA, but not that it is true. In short, as far as van Fraassen is concerned, explanation requires belief in empirical adequacy, not belief in truth.

Let me further flesh out empiricists' position. I claimed in Sect. 7.2.2 of this chapter that a fundamental difference between scientific and fictional understanding is that belief is an ingredient of scientific understanding, but not an ingredient of fictional understanding. On this account, when explainers give scientific explanations, they believe that the explananda and explanantia are true, but when they give fictional explanations, they do not believe that the explananda and explanantia are true. Contrary to this account, empiricists might suggest that a belief in empirical adequacy is an ingredient of scientific understanding, but not an ingredient of fictional understanding. On their account, when explainers give scientific explanations, they believe that the explananda and explanantia are EA, but when they give fictional explanations, they do not believe that the explananda and explanantia are EA. Therefore, empiricists can scientifically explain explananda in terms of explanantia. So might empiricists claim.

In my view, however, empiricists cannot adequately explain events in terms of T. Suppose that they believe that the dragon hypothesis is EA and that they say, "A lunar eclipse occurs because a dragon swallows the moon." When they say so, they are saying more than what they believe. After all, the assertion that the dragon hypothesis is EA does not entail that dragons are real. As a result, empiricists are susceptible to the previously mentioned three problems. Suppose that they tell their interlocutors, "A lunar eclipse occurs because a dragon swallows the moon, but I don't believe dragons are real." Then, Moore's paradox comes to the fore, and their speech act only puzzles their interlocutors. In sum, empiricists cannot escape from the three problems by believing that T is EA, as we noted in Sections 9 and 10 of Chap. 3. Because of these three problems, empiricists cannot help their interlocutors understand events in terms of T.

Moreover, empiricists themselves cannot understand events in terms of T (Park, 2014, p. 287). Imagine that you firmly believe that a lunar eclipse occurs due to the alignment of the three objects and that you are transported in a time machine to ancient China. The ancient Chinese accumulated such extensive data about lunar eclipses that their predictions are all true. Consequently, they believe that the dragon hypothesis is EA. Impressed by their data, you also believe that the dragon hypothesis is EA. However, modern astronomy is so deeply entrenched in your mind that you would resist the dragon hypothesis. Consequently, you would not be able to understand a lunar eclipse in terms of the dragon hypothesis. Instead, you would only think that ancient Chinese misunderstand why a lunar eclipse occurs. In sum, empiricists cannot understand events in terms of T.

The preceding discussion shows that empiricists cannot understand an observational event in terms of a theoretical event. I admit, though, that they can understand an observational event in terms of another observational event. Suppose, for example, that the engine of your car suddenly stops while you are driving, and you do not understand why it has stopped. You call a mechanic, who inspects your vehicle and tells you that there is not enough engine oil. You now understand why your car stopped. Empiricists can believe this explanandum and explanans. Hence, they can understand one observational event in terms of the other. This kind of understanding might be called "observational understanding." Observational understanding arises when an observational event is explained in terms of another. Even our ancestors from the Stone Age, who had not yet developed science, could have had such an understanding.

Observational understanding is to be contrasted with *theoretical understanding*. Theoretical understanding arises when the explanandum and/or the explanans are theoretical events. For example, a theoretical understanding results from the theoretical explanation that the earth contracts and expands because gravitational waves pass through it or from the theoretical explanation that a particular region of spacetime contracts and expands because two massive rocks collide with each other. This kind of understanding is not available for empiricists because they do not believe that theoretical events are real. Our Stone Age ancestors could not have this kind of understanding either because they did not believe that theoretical events were real.

There can be an observational explanation and a theoretical explanation of the same observational event. For example, it is an observational explanation that someone got skin cancer because her skin was often exposed to the sun, and it is a theoretical explanation that she got skin cancer because her skin's DNA was destroyed by ultraviolet rays. The former and the latter give rise to observational and theoretical understanding, respectively. Realists can have both observational and theoretical understanding, whereas empiricists can only have observational understanding.

## 7.3 No Justification?

### *7.3.1 The Episode of Einstein*

Dellsén claims that understanding does not require justification and that it instead requires explanation or prediction. Thus, we can understand explananda in terms of explanantia, although we are not justified in believing them:

> [T]here is no requirement in (U) that an agent has epistemic justification for the propositions on which her understanding is based. More precisely, an agent may partially understand something by virtue of grasping how to correctly explain or predict some aspect of it, even though she does not have the epistemic justification required for knowing the propositions to which she appeals in her explanations/predictions. (Dellsén, 2016, p. 76)

For example, Albert Einstein understood why Brownian motion occurred, using the kinetic theory, although he was skeptical both that Brownian motion was real and that the kinetic theory was true. To justify this claim, Dellsén (2016, p. 76) cites Einstein's following remark:

> In this paper it will be shown that according to the molecular-kinetic theory of heat, bodies of microscopically-visible size suspended in a liquid will perform movements of such magnitude that they can be easily observed in a microscope, on account of the molecular motions of heat. It is possible that the movements to be discussed here are identical with the so-called "Brownian molecular motion"; however, the information available to me regarding the latter is so lacking in precision, that I can form no judgment in the matter. (Einstein, 1905/1956, p. 1)

Dellsén takes this passage as indicating that although an explanandum is unwarranted, we can understand it in terms of an explanans. Dellsén adds that "the kinetic theory of heat was very much up for debate at the turn of the 20th century, with many physicists favoring alternative theories of heat that did not assume the existence of submicroscopic molecules" (Dellsén, 2016, p. 76). Dellsén takes the shaky status of the kinetic theory at the turn of the twentieth century as an indication that although an explanans is unwarranted, we can understand an explanandum in terms of it.

Does this historical episode support Dellsén's thesis that understanding does not require that the explanandum be warranted? My answer is "No." Upon close examination, the quoted passage above indicates that Einstein was sure that the motion of the tiny bodies (the explanandum) was real, although he was not sure that the motion of the tiny bodies was identical with Brownian motion. If the explanandum had been unwarranted, i.e., if Einstein had not had sufficient evidence for the belief that the motion of the tiny bodies was real, he could not have said that the tiny bodies moved around because they collided with molecules.

Let me now turn to the issue of whether the historical episode supports Dellsén's thesis that understanding does not require that the explanans be warranted. The passage cited above indicates that Einstein did not *give* an explanation. Instead, he *reported* an explanation. There is a significant difference between giving and reporting explanations. For example, you are giving an explanation of the motion of the tiny bodies when you say, "The tiny bodies move around because they collide with molecules." By contrast, you are reporting an explanation when you say, "According to the kinetic theory, the tiny bodies move around because they collide with molecules," or when you say, "The collision with molecules explains the motion of the tiny bodies." Einstein uses the latter style of speaking in the passage above.

Why is this distinction between giving and reporting an explanation important? When you give an explanation, you give the impression to your explainees that you believe the explanation. By contrast, when you report an explanation, you give the impression to your explainees that you might not believe the explanation. Consequently, giving an explanation in terms of T is not available for antirealists, while reporting an explanation in terms of T is available for them (Park, 2014, pp. 282–283). Failing to distinguish between giving and reporting an explanation might result in a straw man fallacy. Imagine that you describe to me the ancient Chinese explanation that a lunar eclipse occurs because a dragon swallows the moon. Specifically, you say to me, "According to the ancient Chinese hypothesis, a lunar eclipse occurs because a dragon swallows the moon." To your surprise, however, I accuse you of holding the primitive belief that a dragon swallows the moon. You would quickly reply, "I merely reported the ancient Chinese explanation."

Relatedly, Moore's paradox arises when antirealists give an explanation, but not when they report an explanation. For example, Moore's paradox comes about when antirealists say, "The tiny bodies move around because they collide with molecules, but I don't believe that the tiny bodies move around because they collide with molecules." By contrast, Moore's paradox does not arise when antirealists say, "According to the kinetic theory, the tiny bodies move around because they collide with molecules, but I don't believe that the tiny bodies move around because they collide with molecules." Therefore, antirealists can report explanations that they do not believe.

What follows from the thesis that an explainer should believe an explanation while a reporter or a describer need not? The explanation gives rise to an understanding in the explainer's mind, but it might not in the reporter's mind. Imagine again that you describe to me the ancient Chinese explanation that a lunar eclipse occurs because a dragon swallows the moon and that to your surprise, I accuse you of having misunderstood lunar eclipses. You would protest that you merely reported the ancient Chinese explanation, so it is wrong for me to charge you with having misunderstood lunar eclipses. Your protest is reasonable, which implies that describing an explanation does not necessarily result in an understanding. You cannot understand something by merely describing an explanation of it. If you acquired the understanding when you described the explanation, that is in part because you have endorsed it as well. If you do not endorse it, no understanding arises from describing it. If you merely report others' explanation, you merely report their understanding. To report the understanding is one thing, and to have the understanding is another.

Did Einstein understand the motion of tiny bodies in terms of the kinetic theory when he described the explanation? It is not clear what the correct answer to this question is. If it were clear that he believed the kinetic theory, we would infer that he understood the motion of tiny bodies in terms of the kinetic theory. As the passage cited above indicates, however, he described rather than gave the explanation of the motion of tiny bodies in terms of the kinetic theory. Consequently, we cannot infer that he believed the kinetic theory, nor can we conclude that he understood the motion of tiny bodies in terms of the kinetic theory.

It is worth noting that neither Einstein's remarks above nor the historical fact that the kinetic theory was not established at the turn of the twentieth century demonstrates that Einstein did not believe the kinetic theory. It is simply an open question whether he believed the kinetic theory and hence whether he understood the motion of tiny bodies in terms of the kinetic theory. If he understood the motion of tiny bodies with the theoretical resources from the kinetic theory, that was in part because he *believed* the kinetic theory. Without believing the kinetic theory, he could not have understood the motion of tiny bodies in terms of the kinetic theory.

Does the historical episode of Einstein show that understanding does not require that the explanans be warranted? Einstein did not give the explanation. Instead, he reported the explanation. Therefore, a question does not even arise as to whether he understood the motion of tiny bodies in terms of an unjustified theory, pace Dellsén.

### 7.3.2 *Explanation or Prediction*

As noted in Sect. 7.3.1 above, Dellsén claims that understanding requires explanation or prediction rather than justification. This section unfolds a problem that arises when scientists explain or predict events without justifiably believing that the events in question are real.

Suppose that scientists have developed a new medication and have given it to an experimental group and a control group. They gather experimental data from the two groups. They hope that the new medicine will cure many people. They write an article and submit it to a journal. They state in it that they are not justified in believing their data, but they do *accept* them to explain them in terms of their hypothesis. As a result, they understand the data in terms of the hypothesis. They also state that readers will also be able to understand the data in terms of the hypothesis, provided that they accept the data and the hypothesis.

Such an article would be subject to a desk reject. The editors, if they were not in a forgiving mood, would write to the authors that their paper would be better suited for science-fiction publishers than for academic journals. The editors, if even more unkind, would accuse the authors of having fabricated the data (Park, 2017b, pp. 572–573). Let me add that if the editors were epistemic reciprocalists, they would provide the authors with the explanation that the authors fabricated the data because they were of bad character. Of course, the authors would protest that before giving such an explanation, the editors should first be justified in believing that the authors have fabricated the data. However, the editors would retort that they can give the explanation even though they are not justified in believing that the authors have fabricated the data. They only need to accept the explanandum for the sake of giving the explanation.

Realizing that this absurd consequence follows from the proposal that an explanandum does not need to be justified, Dellsén might now propose that an explanandum should indeed be justified. However, he might insist that an explanans does not need to be justified. Thus, the scientists in the example above should be justified in believing the data that they collected, but they do not need to be justified in believing their hypothesis.

This new proposal, although it improves upon the original proposal, is untenable. In science, a theoretical event is sometimes explained in terms of another theoretical event. Physicists say, for example, that the spacetime around the earth contracts and expands because gravitational waves pass through it (Park, 2020, p. 181). In addition, a theoretical event is sometimes explained in terms of an observational event in science. For example, physicists say that a particular region of spacetime contracts and expands because two massive rocks collide. We can be more confident about the explanans than about the explanandum. Consequently, it is problematic to say that we should be justified in believing the explanandum, but we do not need to be justified in believing the explanans (Park, 2020, p. 181).

Finally, Dellsén might propose that observational statements ought to be warranted while theoretical statements do not need to be, whether observational and theoretical statements serve as explananda or explanantia. However, even this possible move is problematic. Suppose that an observational proposition is justified and that it is a part of an understanding. To say that it is justified implies that it is *believed* and that it is likely to be *true*. After all, the notion of justification has no meaning without the notions of belief and truth. In other words, it does not make sense to say that a statement is justified, but no one believes it, and we cannot even talk about whether it is true. Consider what reliabilists in epistemology say about the notion of justification. They say, "Justification is conferred on a belief by the truth-ratio (reliability) of the process that generates it" (Goldman & Beddor, 2016). Note that the notion of justification is explicated in terms of the notions of belief and truth, i.e., that justification is parasitic on belief and truth. Thus, to say that a justified observational statement is a component of understanding is to admit not only that justification is a component of understanding but also that belief and truth are components of understanding, which is precisely what epistemism claims.

## 7.4 No Truth?

### *7.4.1 Models*

As noted in Sect. 7.1 of this chapter, epistemism would be rejected by Elgin (2004, 2012) and de Regt (2015), who argue, respectively, that models and past theories, although false, nevertheless produce scientific understanding. This subsection and the next critically examine Elgin's argument and de Regt's argument, respectively.

Epistemism makes the following claims about understanding, misunderstanding, and approximate understanding. We understand explananda in terms of explanantia, provided that they are true and we justifiably believe them. We misunderstand explananda in terms of explanantia when explananda and/or explanantia are false but we believe them. We *approximately* understand explananda in terms of explanantia, provided that explananda and/or explanantia are approximately true, but we believe that they are true. Epistemism recognizes that misunderstanding comes in degrees, i.e., that misunderstanding can be close to or far from understanding, depending on whether explananda and/or explanantia are close to or far from truths.

Elgin observes that current science "develops and deploys simplified models that diverge, sometimes considerably, from the phenomena they purport to represent" (Elgin, 2004, p. 113). For example, the kinetic model of gas holds that a gas "consists of dimensionless spherical molecules that are not subject to friction and exhibit no mutual attraction" (Elgin, 2012, p. 136). A model, although false, generates scientific understanding. Therefore, it is wrong to say that an explanans needs to be true to create scientific understanding.

My response to this argument is to argue that a model does not generate understanding but rather misunderstanding. The kinetic model claims that the temperature of a gas is high because the spherical molecules of the gas are moving quickly and that the temperature of another gas is low because the spherical molecules of the gas are moving slowly. If people claim that ice is cold because the spherical ice molecules move slowly, we would reply that they have misunderstood why ice is cold. After all, molecules are not spherical. However, their misunderstanding is close to an understanding, i.e., their misunderstanding is an *approximate* understanding.

What are the reasons for thinking that a model generates approximate understanding? To answer this question, we should engage Elgin's (2012, p. 139) account of models. She claims that a model represents a target system in the way that detergent samples represent the detergents on the market. Detergent manufacturers distribute free samples to potential customers. The samples and the detergents on the market share some, but not all, properties. For example, they share the capability to remove stains, but they do not share the price. The samples are free, but the detergents in the store are not. Thus, some properties of samples are attributable to the target objects, while other properties are not.

Likewise, the kinetic theory treats a gas as if it "consists of dimensionless spherical molecules that are not subject to friction and exhibit no mutual attraction" (Elgin, 2012, p. 136). These properties of the kinetic model are not attributable to gases in the world since molecules in the world are not spherical, and they are subject to mutual attraction and friction. However, the model has other properties that can be attributed to gases. For example, the mass of a gas is the total mass of its molecules, and its temperature is the mean kinetic energy of its molecules. These properties can be found both in the model and in the target system.

Elgin's preceding two examples illustrate the following two important things. First, some of the properties of models can be attributed to target systems, while other properties cannot be. Second, model-users know which properties are attributable and unattributable to target systems. They know, for example, that detergents in the store can remove stains just like the detergent samples, but that they are not free. Scientists likewise know that the mass of a real gas is the total mass of its molecules and that these molecules are not spherical. Hence, model-users are not terribly wrong about target systems. That is, their beliefs about target systems are not completely false; instead, they are approximately true. Therefore, a model generates approximate understanding as opposed to complete misunderstanding.

This conclusion can be augmented by Ronald Giere's (2006, pp. 81–82) comparison of a model to a subway map. A subway map does not precisely represent the subway system. It distorts the distances between the stations and ignores the depth of the stations. However, it does capture the connections between train lines and the order of the stations. Such a subway map is thus more useful to people navigating a city than a more accurate subway map that is filled with complicated information that a subway rider does not need. De Regt endorses Giere's analogy of a model to a subway map, saying that the "way in which scientific theories and models provide understanding is comparable to the way maps help us to move around in the world" (de Regt, 2015, p. 3791).

Note that the subway map correctly represents the connections between the lines and the order of stations, though it incorrectly represents the distances between the stations and ignores the depth of the stations. Moreover, map-users know which properties of the map correspond to the subway system and which properties do not. It follows that the map-users are not terribly wrong about the subway system, i.e., that the map generates approximately true beliefs about the subway system. Given that de Regt endorses Giere's analogy, he should also agree that a model in science generates approximate understanding.

My opponents might raise the following objection to my preceding observation about models. Some models in science distort their targets so wildly that they cannot even be regarded as approximately true, but instead must be considered as completely false. It is thus wrong to say that such a model generates approximate understanding.

This objection runs counter to the idea underlying the analogies of a model to a detergent and a subway map. Sample detergents would fail to satisfy customers if they were profoundly different from the detergents on the market, which would imply that they do not have the capability to remove stains. Subway maps would provoke storms of protest from subway users if they profoundly misrepresent a subway system, which would imply that they misrepresent even the connections between the lines and the order of the stations.

Epistemists can grant that an utterly false model can be useful for predicting and controlling phenomena, but they would point out that the existence of such a model is not a strike against epistemism at all. Epistemists are happy to say that such a model generates complete misunderstanding. If it is legitimate for antirealists to say that a completely false theory can be successful, it should also be legitimate for epistemists to say that complete misunderstanding may result from a successful theory. Suppose, for example, that the kinetic model is completely false, but that it enables us to make quantitatively true predictions about the volume, pressure, and temperature of a gas in a container. In this case, epistemists would say that complete misunderstanding results from a successful model.

## 7.4.2 *Past Theories*

Another objection to epistemism comes from the pessimistic induction (PI), as de Regt notes:

> If one accepts the pessimistic induction, one has to conclude that not only many past theories but plausibly also our current best theories are false. If understanding requires (approximate) truth, this entails that we do not have any explanatory understanding at all. (de Regt, 2015, p. 3790)

It is wrong to assert that phlogiston theorists "like Stahl and Priestley did not possess any scientific understanding at all" (de Regt, 2015, p. 3792). Nor is it right to assert that oxygen theorists today do not possess any explanatory understanding, on the grounds that the oxygen theory will eventually follow the fate of the phlogiston theory. However, epistemism implies that past scientists did not, and present scientists do not, have any understanding at all. Therefore, epistemism is false.

How would epistemists respond to this charge? They would argue that past scientists misunderstood phenomena. For example, phlogiston theorists thought that when an object burned, it became lighter because the phlogiston was released from it. Obviously, they misunderstood why a burning object becomes lighter. To take another example, caloric theorists thought that when a cold object touched a hot object, caloric fluids flowed from the hot to the cold object and that as a result, they reached the average temperature between them. Obviously, they misunderstood why the two objects arrived at the average temperature. In short, past theories produced misunderstanding of what they explained.

However, epistemists might argue that past scientists' misunderstanding was close to understanding, i.e., that their misunderstanding was *approximate* understanding. This contention is based on selective realism, according to which some theoretical components of past theories fueled the success of the theories while other components merely set up the metaphysical frameworks for research. The former survived scientific revolutions, while the latter did not. More importantly, past scientists believed the former, but not the latter. Based on this distinction, Stathis Psillos (1999, p. 113) contends that obsolete theories, although false, were approximately true. Appealing to Psillos's position, epistemists can say that approximate understanding resulted from past theories.

Kyle Stanford might object that past theories were "profoundly mistaken in their most fundamental claims about nature" (Stanford, 2006, p. 8). For example, the phlogiston theory claims that phlogiston is released when an object burns. By contrast, the oxygen theory claims that oxygen is absorbed when an object burns. How can the phlogiston theory be approximately true when it is radically distinct from the oxygen theory? It is a matter of taste whether to apply "approximately true" to past theories (Stanford, 2015, p. 876). Therefore, he would say that it is a matter of taste to say whether past theories produce approximate understanding or complete misunderstanding.

Epistemists can cheerfully accept this objection, given that epistemism does not require that past theories produced approximate understanding. Epistemism only requires that *if* a theory is approximately true, it gives rise to approximate understanding. What if a theory is completely false? Epistemism implies that it gives rise to complete misunderstanding. Therefore, it is a straw man fallacy to accuse epistemism of having the controversial consequence that past theories generate approximate understanding.

Moreover, epistemists could argue that (approximate) understanding results from present theories, appealing to my criticisms against the PI in Chap. 2 and the six new arguments for realism in Chap. 4. They might appeal, for example, to the counter-induction for realism, saying that since complete misunderstanding resulted from past theories, (approximate) understanding results from present theories. Therefore, the PI is a flimsy basis for rejecting my claim that (approximate) understanding results from present theories.

A different sort of objection can be raised against those who say that present theories are utterly false, but that scientists have understood unobservables. If they have the nerve to claim that current theories are utterly false, they should also have the nerve to claim that current scientists utterly misunderstood unobservables. If they hate to say that current scientists completely misunderstood unobservables, they should also hate to say that current theories are completely false. It is a double standard to be willing to assert that current theories are radically false, but to be unwilling to assert that current scientists radically misunderstood unobservables. In my view, if current scientists are *profoundly mistaken* about the world, as Stanford (2006, p. 8) claims, they have *profoundly misunderstood* the world.

Let me turn to a different issue. Past theories are used in the literature to motivate not only the PI but also the Humean argument, as we saw in Sect. 2 of Chap. 2. The Humean argument holds that since past theories were false, their current successors are unwarranted. Antirealists who run this argument against realism might challenge epistemists with the question, "What if current theories are unwarranted? Do they generate understanding or misunderstanding?"

Epistemists' natural answer to this question is that we do not know whether current theories generate understanding or misunderstanding. This position competes with an alternative answer that we do know that they give rise to understanding. Defenders of the alternative are saddled with the burden of giving an account of how understanding and misunderstanding differ from each other and an account of the condition under which misunderstanding arises. This objection will become clearer in the next subsection.

### *7.4.3 Anti-epistemism*

Suppose that anti-epistemists believe that T is false on the grounds that its predecessor was false, it invokes idealized objects, and so forth. They observe, however, that T adequately explains or predicts phenomena. On the basis of this observation, they suggest that understanding requires explanation or prediction, but none of the three ingredients of knowledge: belief, justification, and truth. Let me make a couple of critical comments on anti-epistemism in this subsection.

Anti-epistemists face a challenging task, viz., to provide an account of scientific explanation and prediction without invoking the three ingredients of knowledge: belief, justification, and truth. If they smuggle truth into their account of scientific explanation and prediction, epistemists would quickly point out, "While truth is frequently dismissed sternly at the front door, it often seems to be warmly received at the back" (Kitcher, 1993, p. 130). The same holds for belief and justification.

Carl Hempel's (1966) account of scientific explanation and prediction does not help anti-epistemists because it makes use of all three ingredients of knowledge.

Hempel writes, "the explanatory information adduced affords good grounds for believing that the phenomenon to be explained did, or does, indeed occur" (Hempel, 1966, p. 48). In other words, scientific explanation meets the requirement that explanantia should provide good *grounds* for *believing* that an explanandum occurs. To believe a proposition is to feel that it is *true*. Accordingly, all three ingredients are employed in Hempel's account of scientific explanation. How about scientific prediction? According to Hempel, the only difference between explanation and prediction is that in an explanation, an inference is made from an observable event to T, while in a prediction, an inference is made from T to an observable event. Thus, all three ingredients are also employed in Hempel's account of scientific prediction.

Let me make another critical comment against anti-epistemism. Anti-epistemists claim that T gives rise to understanding, whether it is true or false, as long as it explains and predicts observables. By contrast, epistemists claim that T gives rise to understanding when it is true and misunderstanding when it is false. A perverse question arises for anti-epistemists. When does misunderstanding arise? They might answer that misunderstanding arises when T fails to explain or predict observables. An obvious flaw with this answer is that if T fails to explain or predict observables, we usually throw it out. Once it is thrown out, it does not even have a chance to have an impact on our minds, which implies that it cannot even generate misunderstanding.

I claim that if current theories are profoundly mistaken about unobservables, then current scientists have profoundly misunderstood unobservables, and that if current theories are (approximately) true, then current scientists have (approximately) understood unobservables. To go further, I claim that current theories are (approximately) true. This claim is predicated on what I did in the previous chapters. In Chap. 1, I replied to several standard objections to the no-miracles argument. In Chap. 2, I argued that the PI is infinitely problematic. In Chap. 3, I argued that antirealism has epistemic and pragmatic disadvantages. In Chap. 4, I constructed the six new arguments for realism.

## 7.5 Summary

Dellsén argues that we can understand explananda in terms of explanantia, even though we cannot justifiably believe them, provided that we accept them. I object that such an understanding is no different from a fictional understanding. Empiricists, who believe that T is EA, can have an observational understanding, but not a theoretical understanding. Only realists can have the latter. This difference between realists and empiricists motivates the slogan: "I believe more. Therefore, I understand more."

Elgin and de Regt argue that models and past theories, although false, generate understanding. I object that models and false theories instead generate misunderstanding simply because they are false, but that the misunderstanding is an approximate understanding, provided that models and past theories are approximately true.

If current theories are utterly false, current scientists utterly misunderstood unobservables. However, we have good epistemic and pragmatic reasons for believing that current theories are not utterly false.

Anti-epistemists claim that scientific understanding requires explanation and prediction, but none of the three ingredients of knowledge: belief, justification, and truth. I object that they have the burden of giving an account of explanation and prediction without appealing to the three ingredients and an account of when misunderstanding arises.

## References

Cohen, J. (1992). *An essay on belief and acceptance*. Oxford University Press.

de Regt, H. (2015). Scientific understanding: Truth or dare? *Synthese, 192*(12), 3781–3797.

Dellsén, F. (2016). Scientific progress: Knowledge vs. understanding. *Studies in History and Philosophy of Science, 56*, 72–83.

Dellsén, F. (2017a). Understanding without justification or belief. *Ratio: An International Journal of Analytic Philosophy, 30*(3), 239–254.

Dellsén, F. (2017b). Reconstructed empiricism. *Acta Analytica, 32*(1), 95–113.

Einstein, A. (1905/1956). *Investigations on the theory of the Brownian movement*. Trans. A. Cowper. Dover.

Elgin, C. (2004). True enough. *Philosophical Issues, 14*(1), 113–131.

Elgin, C. (2012). Understanding's tethers. In C. Jäger & W. Löffler (Eds.), *Epistemology: Contexts, values, disagreement* (pp. 131–146). Ontos Verlag.

Giere, R. (2006). *Scientific perspectivism*. University of Chicago Press.

Goldman, A., & Beddor, B. (2016). Reliabilist epistemology. In E. N. Zalta (Ed.), *The Stanford encyclopedia of philosophy*. https://plato.stanford.edu/archives/win2016/entries/reliabilism/

Grimm, S. (2006). Is understanding a species of knowledge? *The British Journal for the Philosophy of Science, 57*(3), 515–535.

Grimm, S. (2014). Understanding as knowledge of causes. In A. Fairweather (Ed.), *Virtue epistemology naturalized: Bridges between virtue epistemology and philosophy of science* (Synthese library) (Vol. 366, pp. 329–345). Springer.

Hempel, C. (1966). *Philosophy of natural science*. Prentice-Hall.

Kitcher, P. (1993). *The advancement of science: Science without legend, objectivity without illusions*. Oxford University Press.

Park, S. (2014). The doxastic requirement of scientific explanation and understanding. *PRO, 13*(2), 279–290.

Park, S. (2016). Scientific realism and antirealism in science education. *Coactivity: Philosophy, Communication, 24*(1), 72–81.

Park, S. (2017a). Understanding without justification and belief? *Principia: An International Journal of Epistemology, 21*(3), 379–389.

Park, S. (2017b). Does scientific progress consist in increasing knowledge or understanding? *Journal for General Philosophy of Science, 48*(4), 569–579.

Park, S. (2020). Scientific understanding, fictional understanding, and scientific progress. *Journal for General Philosophy of Science, 51*(1), 173–184.

Psillos, S. (1999). *Scientific realism: How science tracks truth*. Routledge.
Stanford, K. (2006). *Exceeding our grasp: Science, history, and the problem of unconceived alternatives*. Oxford University Press.
Stanford, K. (2015). Catastrophism, uniformitarianism, and a scientific realism debate that makes a difference. *Philosophy of Science, 82*(5), 867–878.
van Fraassen, B. (1980). *The scientific image*. Oxford University Press.

# Chapter 8
# Scientific Realism and Scientific Progress

**Abstract** There are four competing accounts of scientific progress in the literature. They are the problem-solving, semantic, epistemic, and noetic accounts. I defend a new account of scientific progress that I call "the evidential account." It asserts that science progresses when the amount of evidence for our beliefs about the world increases. I argue that the historical episode of Semmelweis accords well with the epistemic and evidential accounts, but not with the problem-solving, semantic, or noetic accounts of scientific progress. I also explore how each of the five rival accounts of scientific progress relates to realism and empiricism.

**Keywords** Empiricism · Epistemic account · Evidential account · Noetic account · Problem-solving account · Realism · Scientific progress · Semantic account

## 8.1 Introduction

There are four competing accounts of scientific progress. They are the problem-solving account (Kuhn, 1962/1970; Laudan, 1977, 1981, 1984), the semantic account (Niiniluoto, 1984, 2014), the epistemic account (Bird, 2007), and the noetic account (Dellsén, 2016, 2017). They hold, respectively, that scientific progress happens when there is an increase in problem-solving effectiveness, verisimilitude, knowledge, and understanding. They offer different explanantia to explain why an episode in science is progressive or regressive. Finnur Dellsén (2018) nicely summarizes these four rival accounts. This chapter defends a new account of scientific progress that I call "the evidential account" and then explores how the five rival accounts of scientific progress relate to realism and empiricism.

There are cognitive and noncognitive episodes in science. For example, scientists' producing new experimental data is a cognitive episode, whereas scientists' securing research funding or maintaining a good relationship with their colleagues is a noncognitive episode. Noncognitive episodes lie outside of any account of scientific progress (Park, 2020, pp. 174–175). Ilkka Niiniluoto (2015) and Dellsén (2016) would agree with me on this count. Dellsén approvingly cites Niiniluoto


S. Park, *Embracing Scientific Realism*, Synthese Library 445,
https://doi.org/10.1007/978-3-030-87813-9_8

(2015), saying that "accounts of scientific progress concern the kind of cognitive progress that roughly consists in improving the ways in which science represents the world" (Dellsén, 2016, p. 76). Therefore, it is wrong to ask theorists of scientific progress to determine whether a noncognitive episode is progressive or regressive.

The plan for the present chapter is as follows. In Sect. 8.2, I recount the historical episode of Ignaz Semmelweis (1818–1865), against which the five rival accounts of scientific progress will be tested in the subsequent sections. In Sects. 8.3, 8.4, and 8.5, I explicate the problem-solving, semantic, and noetic accounts, respectively. All three of these accounts have the disagreeable implication that Semmelweis did not make scientific progress when he tested and discarded five unsuccessful false hypotheses. In Sect. 8.6, I show that the epistemic account has the agreeable implication that Semmelweis made scientific progress. In Sect. 8.7, I expound the evidential account. It is distinct from all four existing accounts, and like the epistemic account, it has the agreeable implication that Semmelweis made scientific progress. In Sect. 8.8, I argue that the problem-solving and noetic accounts are incompatible with realism and empiricism, while the semantic, epistemic, and evidential accounts are compatible with them, and that if proponents of the three accounts endorse realism and empiricism, they would be claiming more than their accounts of scientific progress require them to claim.

## 8.2 The Historical Episode of Semmelweis

In 1846, many women delivering babies died from so-called childbed fever in the Vienna General Hospital, where there were two maternity wards. The mortality rates of the women in the first and second wards were approximately 10% and 2%, respectively. Semmelweis was puzzled by the excessively high mortality rate of the first ward. He tested the following six hypotheses one by one until he ended the scourge (Hempel, 1966, pp. 3–6).

The first hypothesis was that childbed fever was caused by miasma that spread through the air. Semmelweis quickly rejected this hypothesis on the grounds that it was not compatible with the fact that the mortality rate of the women in the first ward was much higher than the mortality rate of the women in the second ward and the mortality rate of the women who gave birth *en route* to the hospital. If this hypothesis had been true, then the morality rate of the three groups of women should have been about the same.

The second hypothesis was that the first ward was overcrowded. He dismissed this hypothesis on the grounds that the second ward was more crowded than the first ward. In fact, the second ward was more crowded because women were afraid to be sent to the more deadly first ward; some women begged on their knees not to be sent there.

The third hypothesis was that the medical students who treated the women in the first ward were rough with them and caused injuries to them. He ruled out this hypothesis on the grounds that the midwives who treated the women in the second ward were just as rough with them as the medical students were with the women in the first ward and on the grounds that the injuries resulting from natural birth were at least as severe as the injuries that resulted from the students' rough treatment.

The fourth hypothesis was that the sight of a priest and his attendant was the cause of the higher mortality rate. The priest and his attendant walked past the first ward while ringing a bell and heading to a room where they held a religious ritual for a dying woman. It was suggested that the sight of them terrified the women in the first ward. Semmelweis eliminated this hypothesis on the grounds that the mortality rate did not decrease even when the priest and his attendant did not walk past the first ward.

The fifth hypothesis was that the women in the first ward delivered babies on their backs, while those in the second ward did so on their sides. He did not consider this hypothesis to be plausible. He nonetheless tested it, introducing the use of the lateral position to the women in the first ward, but the introduction did not make any difference in the mortality rate, and the hypothesis was thrown out.

Finally, an incident occurred that gave Semmelweis a decisive clue to the cause of the high mortality rate. His colleague accidentally got a wound on his finger using a scalpel, which a medical student had used for an autopsy. The colleague died after having symptoms that were similar to those of the women dying from puerperal fever. The medical students were involved in the autopsy before they treated the women in the first ward. By contrast, the midwives were not involved in the autopsy at all. Semmelweis instituted a policy of washing hands with a chlorinated solution. As a result, the mortality rate dropped to about 2%. He finally arrived at the true hypothesis that the cadaverous materials caused the high mortality rate.

This historical episode is cognitive, given that to replace a hypothesis with another one is to replace a representation of the world with another one. Such an episode is different in kind from noncognitive episodes, such as securing research funding or maintaining a good relationship with one's colleagues. Therefore, a theory of scientific progress should tell us whether and why the episode of Semmelweis is progressive or regressive.

This historical episode radically differs from the other historical episodes that philosophers of scientific progress discuss. In the latter episodes, old and new theories are all successful, and they share some common assumptions. For example, the caloric and kinetic theories were successful, and they share the law of conservation of mass. In the episode of Semmelweis, however, the first five hypotheses were unsuccessful, and only the sixth was successful. In addition, the six hypotheses did not share any theoretical assumptions. Consequently, the episode of Semmelweis casts new light on rival accounts of scientific progress, which will become clear in the following sections.

## 8.3 The Problem-Solving Account

According to the problem-solving account, an episode is progressive when more problems or puzzles are solved after the episode. The replacement of a theory with its successor is progressive insofar as the new one solves more problems than the old one. Thus, the replacement of the caloric theory with the kinetic theory was progressive.

The distinctive characteristic of the problem-solving account is that it resists truth. Thomas Kuhn asserts that "the notion of a match between the ontology of a theory and its 'real' counterpart in nature now seems to me illusive in principle" (Kuhn, 1962/1970, p. 206). Larry Laudan states that truth "is intrinsically transcendent and hence closed to epistemic access" (Laudan, 1981, p. 145). Thus, to say that a new theory solves more problems than an old one does not mean that the new theory is more likely to be true or is more truthlike than the old one, nor does it mean that the new theory makes more true predictions than the old one.

Then what does it mean to solve a problem? What exactly are we doing when we solve a problem? It is a challenging task to answer this question. Supporters of the problem-solving account might suggest that to solve a problem is to make *useful* predictions. However, this suggestion invites a further question: what does "useful predictions" mean? They cannot answer that it means *true* predictions because that goes against their aim of giving an account of scientific progress without appealing to truth. Philip Kitcher critically examines a few such attempts and then declares that "While truth is frequently dismissed sternly at the front door, it often seems to be warmly received at the back" (Kitcher, 1993, p. 130).

Kitcher's conceptual objection to the problem-solving account is not a lethal one. Supporters of the problem-solving account can get around Kitcher's objection by appealing to Bas van Fraassen's (1980, p. 16) insight. Van Fraassen's (1980) position requires a distinction between observables and unobservables. Grover Maxwell (1962) objects that we cannot draw a line between observables and unobservables. Van Fraassen replies that the predicate "observables," although vague, is "usable provided it has clear cases and clear counter-cases" (van Fraassen, 1980, p. 16). In van Fraassen's spirit, supporters of the problem-solving account might argue that although there is no precise definition of "problem-solving," the term is usable because there are clear cases of solutions to problems. For example, Semmelweis solved a problem when he brought down the high mortality rate of the first ward.

The problem-solving account has an empirical problem, i.e., it cannot accommodate the episode of Semmelweis. It should imply, but it does not, that his elimination of the five false hypotheses counts as scientific progress. None of the five hypotheses was better than its predecessor in being able to bring down the high mortality rate. In other words, the capacity to solve the problem did not increase as Semmelweis moved from one hypothesis to the next in the series of the five false hypotheses. Therefore, the problem-solving account implies that the elimination of the five hypotheses does not count as scientific progress and that Semmelweis did not make scientific progress until he arrived at the sixth hypothesis.

What is wrong with the implication that eliminating the five false hypotheses does not count as scientific progress? Scientists often have failures before they make successes. For example, scientists failed more than 200 times before they succeeded in cloning Dolly in the 1990s (Park, 2017, p. 216). In science, the road to success is paved with failures. The more failures scientists have, the more likely it is that they will succeed in the next attempt and the closer they are to their destination. To have a failure is to take a step toward success and to make scientific progress. It is thus wrong to think that having a failure does not count as scientific progress and that only achieving success in the end counts as scientific progress.

Let me turn to an alternative to the problem-solving account. As Dellsén (2018, Section 6) observes, one might suggest that scientific progress consists in getting closer to empirical adequacy. This proposal might be called "the empirical account." It differs from the problem-solving account in that it does not eschew truth. After all, to say that T is empirically adequate (EA) means that all of what it says about observables is true. Therefore, the empirical account gets around Kitcher's conceptual objection to the problem-solving account above.

The empirical account cannot accommodate the episode of Semmelweis. None of the five false hypotheses explained why the death rate was inordinately high in the first ward. It follows that Semmelweis was not moving toward empirical adequacy with respect to the high mortality rate. He moved toward empirical adequacy only when he reached the sixth hypothesis. Consequently, the empirical account absurdly implies that he did not make scientific progress until he obtained the sixth hypothesis.

## 8.4 The Semantic Account

The semantic account asserts that "scientific progress can be defined by increasing verisimilitude" (Niiniluoto, 2014, p. 77). On this account, scientific progress occurs when a new theory supplants an old one, insofar as the new one is more verisimilar than the old one, i.e., the later theory is more truthlike than the earlier one. Scientific regress occurs when a new theory is less verisimilar than an old one.

The semantic account does not say that a new theory is truthlike, but says that it is more truthlike than an old one. There is a vast difference between the absolute thesis that a new theory is verisimilar and the relative thesis that a new theory is more verisimilar than an old one, as Brad Wray (2008, p. 323) and Moti Mizrahi (2013, p. 401) observe. A new theory might be more verisimilar than an old one, although both of them are far from being verisimilar. Hence, the semantic account is not refuted even if new theories are not verisimilar.

The semantic account does not say that it is *likely* that a later theory is more truthlike than an earlier one. Instead, it says that a later theory is *definitely* more truthlike than an earlier one. Thus, it can be refuted by a single counterexample that shows that scientific progress is made while there is no increase in verisimilitude. To use an analogy, the view that knowledge is a justified true belief can be refuted by a single counterexample in which a justified true belief is not knowledge.

Philosophers of science think that the discontinuity of reference between an old theory and a new one means that there is no scientific progress. Howard Sankey says, for instance, "If there is discontinuity of reference between theories, there may be no scientific progress in the sense of increase of truth about a common domain of entities" (Sankey, 2009, p. 196). If the theoretical terms of $T_1$ do not refer, but the theoretical terms of $T_2$ refer, or if none of the theoretical terms refer, it is problematic to say that $T_2$ is more truthlike than $T_1$. Therefore, proponents of the semantic account have the burden of developing a theory of reference that would imply that the theoretical terms of successive theories refer to the same theoretical entities.

Proponents of the semantic account might appeal to the theories of reference developed by selective realists, such as Kitcher (1993, pp. 75–105) and Stathis Psillos (1999, 280–300). Consider, for example, the ether theory and the electromagnetic theory:

> The two theories may differ in the full descriptions of the properties they ascribe to the posited causal agents. But there may well be a *substantive overlap* between them. Should this happen, there will be a sense in which there is some referential continuity between the two theories (Psillos, 1999, p. 294).

The ether theory and the electromagnetic theory share what Psillos calls "core causal descriptions," and "luminiferous ether" (approximately) refers to an electromagnetic field (Psillos, 1999, p. 296).

How does the semantic account fare vis-à-vis the historical episode of Semmelweis? The second hypothesis that overcrowding caused childbed fever was no closer to the truth than the first one that miasma was the cause. The fourth hypothesis that the sight of the priest and his attendant was the cause was no closer to the truth than the third one that the medical students' rough treatment was the cause. The fifth hypothesis that having a supine posture during delivery was the cause was no closer to the truth than the fourth one that the sight of the priest and his attendant was the cause. Consequently, the semantic account implies that Semmelweis made no scientific progress until he reached the sixth hypothesis.

The five false hypotheses do not share any noteworthy assumptions. Hence, it is not clear that proponents of the semantic account could develop a theory of reference that would imply that the presumed five causes mentioned in the five false hypotheses were, in fact, the cadaverous materials from the autopsy room. If a theory of reference were to have this implication, we would not think that the putative five causes were, in fact, the cadaverous materials, but rather that the theory of reference is absurd.

## 8.5 The Noetic Account

Dellsén's noetic account asserts that "scientific progress consists in increasing understanding," and "an episode in science is progressive precisely when scientists grasp how to correctly explain or predict more aspects of the world at the end of the episode than at the beginning" (Dellsén, 2016, p. 73). New theories predict or explain more aspects of the world than old ones; as a result, we understand more about the world after scientific revolutions. Dellsén insists that there is no requirement "to the effect that the agents *believes* that the relevant theories are true—she may merely *assume* or *accept* that they are true for the purposes of explanation and/or prediction" (Dellsén, 2016, p. 76).

How does the noetic account relate to the empirical account that was unpacked in Sect. 8.3 of this chapter? There is a similarity between them. Both are based on the common observation that new theories explain or predict more phenomena than old ones. There is a difference between them. They draw different conclusions from the common observation. The noetic account concludes that we understand more aspects of the world than before, whereas the empirical account concludes that we get closer to empirical adequacies.

To remove belief and justification from understanding is to relegate scientific understanding to the level of fictional understanding, as we discussed in Sect. 7.2 of Chap. 7. I set aside this criticism against the noetic account in this chapter. Instead, I test it against the historical episode of Semmelweis. A hypothesis did not correctly explain or predict more aspects of the high mortality rate than its predecessor in the five successive false hypotheses. It follows that according to the noetic account, Semmelweis's understanding of the high mortality rate did not increase and that "Semmelweis was not making any scientific progress when he was eliminating the five false hypotheses" (Park, 2020, p. 175). This verdict of the noetic account is to be contrasted with that of the epistemic account, to which I turn now.

## 8.6 The Epistemic Account

The epistemic account holds that "an episode in science is progressive when at the end of the episode there is more knowledge than at the beginning" (Bird, 2007, p. 92). On this account, the replacement of an old theory with a new one counts as scientific progress because scientists acquire "the negative knowledge that some earlier theory is false" (Bird, 2007, p. 108). Scientists also acquire what might be called "the relative knowledge" that a new theory is more truthlike than an old one. Alexander Bird writes, "The improving precision of our approximations can be an object of knowledge" (Bird, 2007, p. 105).

Let me apply the epistemic account to $T_1$, $T_2$, and $T_3$, where $T_1$ and $T_2$ are obsolete theories while $T_3$ is a current theory. According to the epistemic account, the replacement of $T_1$ with $T_2$ counts as scientific progress, even though they are both false, because scientists of $T_2$ acquired the negative knowledge that $T_1$ was false and because they acquired the relative knowledge that $T_2$ is more verisimilar than $T_1$. What about the replacement of $T_2$ with $T_3$? Bird rejects the pessimistic induction (PI), noting that it is "a poor inference" (Bird, 2007, p. 109). Consequently, he would say that the replacement of $T_2$ with $T_3$ counts as progress because scientists gained the negative knowledge that $T_2$ is false and because they acquired the absolute knowledge that $T_3$ is true.

Can the epistemic account cope with the historical episode of Semmelweis? A positive answer can be inferred from Bird's observation that after a scientific revolution, scientists acquire the negative knowledge that an old theory was false. When Semmelweis tested and discarded the first hypothesis, he formed the negative knowledge that miasma was not the cause of puerperal fever and similarly for the other four false hypotheses. Therefore, Semmelweis was making progress little by little as he eliminated the five false hypotheses one by one.

This implication is different from that of the problem-solving, semantic, and noetic accounts that we discussed in the previous sections of this chapter. These three accounts imply that Semmelweis failed to make scientific progress, because he failed to solve the problem, because the five false hypotheses were not getting any closer to the truth, and because none of the five false hypotheses explained more aspects of the world than its predecessor, respectively.

How can we adjudicate between these two opposing implications? Why is the implication of the epistemic account agreeable? Why is the implication of the other three accounts disagreeable? As we noted in Sect. 8.3 of this chapter, it is wrong to think that having failures on the road to success does not count as progress. Let me present another argument here. Imagine that proponents of the epistemic, problem-solving, semantic, and noetic accounts were with Semmelweis when he had just discarded the fifth hypothesis. They said to him, respectively, the following:

(E) "You have made scientific progress because you now know that the previous five hypotheses were false."
(P) "You have made no scientific progress because none of the five hypotheses was better than its predecessor in terms of the capacity to bring down the high mortality rate."
(S) "You have made no scientific progress because you are no closer to the truth than before."
(N) "You have made no scientific progress because you do not understand more aspects of the high mortality rate than before."

Semmelweis would respond favorably to (E), but not to (P), (S), or (N). Specifically, he would say that (E) has an insight into what he was doing, while (P), (S), and (N) do not. He would add that no progress can be made when philosophers of science advance (P), (S), and (N) to him (Park, 2020, p. 176).

In this context, it is noteworthy that the notion of scientific progress is normative, whereas that of scientific development is descriptive. Niiniluoto states that "'Progress' is an axiological or a normative concept which should be distinguished from neutral descriptive terms like 'change' and 'development'" (Niiniluoto, 2014,

p. 75). Thus, to say that Semmelweis made no scientific progress is to express a negative evaluation of what he did. Consequently, it is natural that Semmelweis would respond positively to (E), but negatively to (P), (S), and (N).

Why should philosophers of science care about how scientists would respond to philosophical theories of scientific progress? This question can be answered with the use of logical positivism as an example. Logical positivism was in vogue in the 1930s, but it had fallen out of fashion by the mid-twentieth century. Imagine that scientists said to philosophers of science in the mid-twentieth century that the demise of logical positivism was not a sign of philosophical progress and that philosophers of science had made no philosophical progress. Philosophers of science would have responded negatively to these scientists, just as Semmelweis would have responded negatively to the authors of (P), (S), and (N) above.

Philosophers of science could have said in the mid-twentieth century that as a result of the demise of logical positivism, they learned that logical positivism was problematic and that they got closer to an unproblematic philosophical theory of science. On this account, philosophers of science learn from previous philosophers' mistakes, and they make philosophical progress through having failures. Thus, the more philosophers have failures, the more likely it is that they will succeed the next time and the closer they get to success.

Just as it would have been reasonable to give such a defense of the demise of logical positivism, so it would have been reasonable to give the epistemic defense of the demise of Semmelweis's five false hypotheses, i.e., to give (E). In general, the demise of a theory, whether scientific or philosophical, constitutes second-order evidence for the truth of its successor. The more false predecessors a theory has, the more likely it is to be true. This observation will be utilized in my defense of a new theory of scientific progress that I call "the evidential account," to which I turn now.

## 8.7 The Evidential Account

### *8.7.1 Content*

According to the evidential account, scientific progress consists in increasing evidence for our beliefs about the world. An episode is progressive if and only if we have more evidence for our beliefs about the world than before, which implies that our beliefs are more likely to be true and/or truthlike than before. The evidential account makes the relative claim, not the absolute claim, that after the episode, our beliefs are likely to be true and/or truthlike. Consequently, the evidential account is not refuted, even if we do not have sufficient evidence for our beliefs. The evidential account originates from the fundamental view about science that like philosophy, it is an evidence-gathering enterprise. Its motto is "Where there is no evidence, there is no science."

Suppose that scientists perform a new experiment or observe a new phenomenon and that as a result, an existing theory is given further support. For example, scientists discovered the fossil of Tiktaalik, the intermediate organism between fish and amphibians (Shubin et al., 2006). The discovery of the fossil reinforced evolutionary theory. The evidential account sees such an episode as a mark of progress because scientists gained new evidence for an existing theory, and as a result, the amount of evidence for our beliefs about the world increased.

Evidence for a theory can come not only from observations but also from neighboring theories. Evolutionary theory supported plate tectonics and vice versa when scientists confirmed their joint prediction that marsupial fossils exist in Antarctica. The evidential account implies that such an episode is progressive. This implication goes hand in hand with the argument from neighboring theories in Sect. 4.4 of Chap. 4 that since T has been augmented by previously unconceived neighboring theories, it will continue to be augmented by hitherto unconceived neighboring theories.

Suppose that scientists perform experiments that disconfirm an old theory but confirm a new one, and that as a result, a scientific revolution occurs. According to the evidential account, a scientific revolution counts as scientific progress as long as we have more evidence for our beliefs about the world than before. There are three reasons for thinking that the volume of evidence for our beliefs about the world increased after a scientific revolution.

First, as many participants in the realism debate have pointed out, a new theory is more successful than an old one, i.e., the former has a broader predictive and explanatory scope than the latter. Even antirealists accept this difference between old and new theories, although they are skeptical that present theories "have achieved a degree of success that warrants their truth" (Müller, 2015, p. 406). To say that a new theory is more successful than an old one means that the amount of observational evidence for the former is greater than the amount of observational evidence for the latter and that we have evidence for the belief that an old theory is false.

Second, evidence for the belief that the old theory is false comes not only from new observational evidence but also from the new theory. The new theory constitutes evidence against an old one. For example, after the chemical revolution, the oxygen theory was evidence against the phlogiston theory. As a result, it was legitimate to cite the oxygen theory as evidence against the belief that a burning object released phlogiston. On this account, a theory can act as evidence for or against another theory. This view about the function of a theory goes hand in hand with the argument from neighboring theories, according to which neighboring theories can support each other.

Third, the collapse of an old theory constitutes second-order evidence for a new one, we noted in Sect. 4.5 of Chap. 4, where the counter-induction for realism was enunciated. According to the counter-induction, the greater the number of false predecessors a theory has, the more likely it is that it is true. Suppose that there are three sequential theories: $T_1$, $T_2$, and $T_3$. The counter-induction implies that $T_3$ is the most likely to be true and $T_1$ is the least likely to be true, which coincides with our intuition. This intuition can be justified by the use of an analogy. Suppose that there are four blue balls and one green ball in a jar. Your task is to pull out the balls one at a time until you pull out the green ball. Under this condition, pulling out a blue ball constitutes second-order evidence for believing that the next ball will be the green one. The same holds for eliminating a false theory from the space of rival theories (Ruhmkorff, 2013, p. 412; Park, 2016, p. 840).

The evidential account does not require that scientists should have all the foregoing three kinds of evidence in order for a scientific revolution to count as scientific progress. In other words, even if one or two kinds of evidence are missing, a scientific revolution can count as scientific progress. On the evidential account, what matters is not whether all three kinds of evidence are present but rather whether the amount of evidence for our beliefs about the world increases. Keep in mind that the acquisition of only one kind of evidence is enough to make a scientific episode progressive.

The evidential account claims that the increase in evidence makes a new theory likely, but not guaranteed, to be more truthlike than an old one. Thus, the evidential account allows for some historical episodes in which an old theory is more verisimilar than a new one. For example, Ptolemy rejected Aristarchus's heliocentric theory and accepted the geocentric theory (Frost-Arnold, 2019, p. 911). In this episode, the old theory was more truthlike than the new one. This episode does not refute the evidential account, insofar as Ptolemy had more evidence for the geocentric theory than for the heliocentric theory. The evidential account would be refuted, though, if such episodes are more frequent than episodes in which new theories are less truthlike than old ones.

The evidential account would also be refuted by a historical episode in which scientific progress is made and yet scientists do not have more evidence after the episode. It is not clear that critics can offer such a counterexample. As noted earlier, a new theory was empirically more successful than an old one in the history of science, which implies that there was more observational evidence for the former than for the latter. In addition, the new theory counts as evidence against the old one, and the demise of the old theory constitutes second-order evidence for the new one. For these three reasons, we can say that the amount of evidence increases when a new theory superseded an old one in the history of science.

Is the historical episode of Semmelweis a counterexample against the evidential account? My answer is "No." The evidential account implies that Semmelweis made scientific progress little by little when he tested and rejected the five false hypotheses one by one. For example, when he tested and rejected the first one, he gained observational evidence for the belief that the miasma was not the cause of the high mortality rate, and he gained second-order evidence for the second hypothesis. The second-order evidence makes the second hypothesis more likely to be true than the first hypothesis. The same holds for the remaining four false hypotheses.

The evidential account can handle the episode of Semmelweis, while the problem-solving, semantic, and noetic accounts cannot, because it imposes a lighter requirement on scientific progress than they. The evidential account does not require that a hypothesis solves more problems than its forerunner, and thus the fact that problem-solving effectiveness did not increase in the five successive false hypotheses is not a blow to the evidential account. The evidential account does not require that a hypothesis should be closer to the truth than its forerunner, and thus the fact that verisimilitude did not increase in the five successive false hypotheses does not refute the evidential account. The evidential account does not require that we understand more aspects of the world than before, and thus the fact that understanding did not increase in the five successive false hypotheses is not a strike against the evidential account. The evidential account only requires that the amount of evidence for our beliefs about the world increases. The increase in the amount of evidence is demonstrated by the increase in the number of rejected theories. Therefore, exponents of the evidential account are more likely to respond favorably to scientific episodes than exponents of the other three accounts.

Relatedly, the evidential account imposes such a light requirement on scientific progress that espousers of the evidential account happily comply with the epistemic imperative: “Act only on an epistemic maxim through which you can at the same time will that it should become a universal one” (Park, 2018, p. 441). They do not have a problem with scientists claiming that their philosophical failures constitute second-order evidence for forthcoming philosophical success. Consequently, they cheerfully will the maxim “Take your epistemic colleagues’ failures as second-order evidence for forthcoming success” to be a universal one.

It is not a demanding request that scientists provide discarded hypotheses to prove that they made scientific progress, but it is a demanding request that scientists provide increasing problem-solving effectiveness, verisimilitude, or understanding to prove that they made scientific progress as the episode of Semmelweis indicates. It is doubtful that espousers of the problem-solving, semantic, and noetic accounts would comply with the epistemic imperative. After all, to comply with the epistemic imperative means that they are willing to provide increasing problem-solving effectiveness, verisimilitude, or understanding to prove that they made philosophical progress.

### 8.7.2 A Comparison with the Problem-Solving Account

How does the evidential account differ from the problem-solving account? They are incompatible with each other in that the problem-solving account resists truth while the evidential account embraces it. On the evidential account, to say that a new theory is supported by more evidence than an old one means that the new theory is more likely to be true and/or truthlike than the old one. Note that truth figures in the formulation of the evidential account.

How does the evidential account relate to the empirical account, according to which scientific progress lies in getting closer to empirical adequacy? The two accounts embrace truth, and they are compatible with each other, but there is an important difference between them. The empirical account requires the *semantic* claim that a new theory is closer to empirical adequacy than an old theory, whereas the evidential account requires the *epistemic* claim that a new theory is supported by more evidence than an old theory. The epistemic thesis can serve as a premise for the semantic thesis, but not vice versa.

The epistemic thesis makes the semantic thesis likely to be true, but not guaranteed to be true. To put it differently, if a new theory is supported by more evidence than an old one, it is likely, but not guaranteed, that the new theory is closer to empirical adequacy than the old one. As the episode of Ptolemy illustrates, even if a theory is more successful than its competitor, it might be farther from empirical adequacy than its competitor. It follows that although the epistemic thesis is true, the semantic thesis might be false and that it is harder to come by the semantic thesis than the epistemic thesis. For this reason, we can say that the evidential account imposes a lighter requirement on scientific progress than does the empirical account.

I earlier claimed in Sect. 8.7.1 of this chapter that when a new theory supersedes an old one, we acquire the three kinds of evidence for our beliefs about the world. I do not rule out the possibility that advocates of the empirical account might endorse my claim, but argue that the acquisition of the three kinds of evidence is an indication that the new theory is closer to empirical adequacy than an old one, so the replacement of the old theory with the new one is progressive. If they make this move, they might be able to handle the episode of Semmelweis. I leave to proponents of the empirical account the task of developing this line of thought.

### 8.7.3 A Comparison with the Semantic Account

The evidential and semantic accounts claim that the replacement of an old theory with a new one is progressive when we have more evidence for the new theory than for the old one and when the new theory is more truthlike than the old one, respectively. Let me highlight the noteworthy differences between these two accounts.

The gathering of evidence is a means to know whether a new theory is more truthlike than an old one. Accordingly, evidence plays profoundly different roles in the semantic and evidential accounts. On the semantic account, the acquisition of evidence is a *means* to know whether science makes progress. On the evidential account, by contrast, the acquisition of evidence is *constitutive* of scientific progress. The gathering of evidence counts as scientific progress, even if an old theory happens to be more verisimilar than a new one.

As a consequence, the semantic and evidential accounts issue different verdicts on the historical episode in which Ptolemy rejected the heliocentric theory and accepted the geocentric one. On the evidential account, the episode was progressive insofar as Ptolemy had more evidence for the geocentric theory than for the heliocentric one. On the semantic account, however, it was regressive, even if Ptolemy had more evidence for the geocentric theory than for the heliocentric one, because the heliocentric theory was more truthlike than the geocentric one.

Another noteworthy difference between the semantic and evidential accounts is that the semantic account requires, whereas the evidential account does not, a theory of reference that would show that the key terms of the successive theories refer to the same theoretical entities although they appear to refer to different ones. Recall that an increasing amount of verisimilitude requires, whereas an increasing amount of evidence does not, the continuity of reference between successive theories.

For these two reasons, we can say that the evidential account imposes a lighter requirement on scientific progress than does the semantic account. Hence, advocates of the evidential account are more likely to respond positively to scientific episodes than advocates of the semantic account. It is natural that unlike the semantic account, the evidential account implies that Semmelweis made scientific progress little by little when he eliminated the five false hypotheses one by one, as we have seen in Sect. 8.7.1 above.

### 8.7.4 A Comparison with the Noetic Account

How does the evidential account differ from the noetic account? They are incompatible with each other. The noetic account resists belief, while the evidential account embraces it. On the evidential account, the increase in evidence means that our beliefs about the world become more probable. How about justification? Both accounts deny that justification is required for scientific progress. However, the evidential account is friendlier to justification than the noetic account, for justification is nothing but the sufficient amount of evidence.

Espousers of the noetic account might withdraw the contention that belief and justification are not required for understanding, and they might only claim that our understanding of the world increases when a new theory explains or predicts more phenomena than an old one. How does this new noetic account relate to the evidential account? Both of them imply that the replacement of the old theory with the new one counts as scientific progress. However, they have the following noteworthy differences.

Supporters of the noetic account and those of the evidential account make different inferences from the same fact that a new theory explains or predicts more phenomena than an old one. Supporters of the noetic account infer that our understanding of the world increased, so scientific progress has been made. By contrast, supporters of the evidential account infer that the amount of evidence for our beliefs about the world increased, so scientific progress has been made. They adduce the different premises for the same conclusion that scientific progress has been made.

Exponents of the evidential account add that there are other reasons for thinking that the amount of evidence increases after a scientific revolution. Scientists have a new theory which is evidence against an old theory and the demise of the old theory which is second-order evidence for the new theory. I do not rule out the possibility that exponents of the noetic account might endorse this claim, but argue that the acquisition of such evidence is an indication that our understanding of the world increased. If they make this move, they might be able to handle the episode of

Semmelweis. I leave to advocates of the noetic account the task of fleshing out this line of thought.

### *8.7.5 A Comparison with the Epistemic Account*

How does the evidential account differ from the epistemic account? The epistemic account provides an increase in knowledge as an explanans for scientific progress. Justification is a component of knowledge, and the degree of justification is proportional to the amount of evidence. Consequently, it is much easier to obtain evidence than knowledge. For example, Ptolemaic scientists had evidence for the proposition that the earth is stationary at the center of the universe, but they did not have the knowledge that the earth is stationary. For these reasons, we can say that the evidential account imposes a far lower requirement on scientific progress than does the epistemic account. Let me go through some scientific episodes to better differentiate between the evidential and epistemic accounts.

Suppose that scientists gather new observational data that support their existing theory. For example, scientists observed gravitational lensing with the Hubble Space Telescope (van der Well et al., 2013). Their observation bolstered general relativity, which was previously confirmed by other observational data. This episode counts as scientific progress under the epistemic account, given that scientists gained new observational *knowledge*. It also counts as scientific progress under the evidential account, given that scientists gained new *evidence* for our beliefs about the world. Therefore, such an episode cannot adjudicate between the evidential and epistemic accounts.

An episode in which an old theory is replaced with a new one cannot adjudicate between the evidential and epistemic accounts either. According to the evidential account, we gain three kinds of new *evidence* for our beliefs about the world in such an episode. They are the greater amount of observational evidence for the new theory, the new theory which is evidence against the old one, and the demise of the old theory which is second-order evidence for the new one. According to the epistemic account, we gain new observational knowledge for the new theory, the negative knowledge that the old theory is false, and the relative knowledge that the new theory is more verisimilar than the old one. Accordingly, a scientific revolution marks progress for both accounts.

The epistemic and evidential accounts might issue different verdicts concerning the historical episode in which Ptolemy rejected Aristarchus's heliocentric theory and accepted the geocentric theory. According to the evidential account, the episode was progressive insofar as Ptolemy had more evidence for the geocentric theory than for the heliocentric theory. What about the epistemic account? Supporters of the epistemic account can give the following two opposing answers.

The first possible answer is that the episode of Ptolemy is regressive. The justification for this answer is that the heliocentric theory makes the true claim that the earth revolves around the sun, whereas the geocentric theory makes the false claim that the sun revolves around the earth. Thus, Ptolemy's choice of the heliocentric theory over the geocentric one involves the loss of the theoretical knowledge that the earth revolves around the sun.

The second possible answer is that the episode of Ptolemy is progressive. The justification for this answer is that the amount of observational evidence for the geocentric theory outweighed the amount of observational evidence for the heliocentric theory, and the gain in observational knowledge was more significant than the loss of the theoretical knowledge that the earth revolves around the sun. As a result, scientists overall gained more knowledge after Ptolemy rejected the heliocentric theory and accepted the geocentric one.

It is an open question whether supporters of the epistemic account would answer in the first or second ways above. Only after they choose one of them can we decide whether the epistemic and evidential accounts issue the same verdict or different verdicts regarding the episode of Ptolemy.

### 8.7.6 Objections and Replies

A reviewer asks the following question about the evidential account. How does it differ from logical positivists' cumulative account that Kuhn (1962/1970) was reacting to? According to the cumulative account, science develops by accumulating new beliefs upon old beliefs, which implies that theories are reduced to their successors. The cumulative account competes with Kuhn's evolutionary account, according to which science develops by replacing old beliefs with new ones, which implies that theories are replaced by their successors.

There are similarities and differences between the cumulative and evidential accounts. A similarity between them is that they capture the cumulative aspect of science. However, they differ on what is accumulated. On the cumulative account, our beliefs are accumulated, as science develops. On the evidential account, by contrast, evidence for our beliefs is accumulated. The evidential account is neutral about whether theories are reduced to or replaced by their successors. Scientific progress is made as long as evidence is accumulated. Therefore, the evidential account is compatible, while the cumulative account is incompatible, with the historical episodes in which theories were replaced by their successors.

There is a more fundamental difference between the cumulative and evidential accounts. The cumulative account is an account of scientific development, whereas the evidential account is an account of scientific progress. "Development" is a descriptive word, whereas "progress" is a normative word, as we noted in Sect. 8.6 of this chapter. The cumulative account is a description of how science changes as time passes, and it is free of a value judgment. By contrast, a value judgment is inherent in the evidential account. According to the evidential account, it was good that Semmelweis tested and rejected the first hypothesis because the rejection of the hypothesis constituted the second-order evidence for the second hypothesis.

The reviewer raises the following objection to the evidential account. There are non-empirical components of scientific progress. This seems to be excluded by the evidential account. Think of Laudan's (1977) notion of conceptual problems or Kitcher's (1993) notion of conceptual progress. Surely there are conceptual aspects of scientific progress. For example, removal of internal inconsistency or ambiguity, introduction of more precise vocabulary, or development of new concepts can all contribute to or constitute progress. Yet, the evidential account of scientific progress appears to grant no place to conceptual aspects of scientific progress.

The reviewer is right to say that conceptual progress is scientific progress. The proponents of the problem-solving, semantic, noetic, epistemic, and evidential accounts have the burden of specifying how their accounts relate to conceptual progress. In my view, they can appeal to Bird's insight that scientific progress is made not only when scientists acquire knowledge but also when they obtain a means to acquire knowledge (Bird, 2007, p. 111). For example, supporters of the evidential account can say that removing internal inconsistency constitutes scientific progress because as a result of removing internal inconsistency, a theory becomes a more convenient device for gathering evidence for our beliefs about the world.

## 8.8 Relationships to Realism and Empiricism

How do the five accounts of scientific progress relate to realism? The five accounts and realism are answers to different questions. The five accounts are answers to the question: what makes a cognitive episode in science progressive or regressive? By contrast, realism is an answer to the question: what epistemic attitude should we take toward our best theories? Although they are answers to different questions, they are nevertheless connected with each other. As Dellsén observes, however, the connection between them has been "underexplored" (Dellsén, 2018, Section 6). This section aims to shed light on this underexplored territory. I explore how the five accounts relate to realism and empiricism, which are defined in this section respectively as the position that T is (approximately) true and the position that T is (approximately) EA. This section is intended to be useful to theorists of scientific progress who wonder whether they should endorse realism and empiricism.

Let me begin with the problem-solving account. The problem-solving account is incompatible with realism and empiricism. After all, the problem-solving account openly eschews truth, whereas both realism and empiricism invoke it. Realism claims that T is true; empiricism claims that it is (approximately) EA, which means that most or all of what T says about observables is true. Consequently, adherents of the problem-solving account should reject both realism and empiricism.

As noted in Sect. 8.3 of this chapter, however, there is an alternative to the problem-solving account, viz., the empirical account, according to which scientific progress occurs when we get closer to empirical adequacy. Suppose that T is closer to empirical adequacy than its predecessor. The empirical account claims that scientific progress was made when T displaced its predecessor. By contrast, realism claims that T is true. Realism entails the relative thesis that T is closer to empirical adequacy than its predecessor. If adherents of the empirical account accept realism, they would be claiming far more than they are required to with respect to T, and they would have to shoulder the burden of diffusing the traditional objections to realism, such as the PI. By contrast, if adherents of the empirical account reject realism, they would not have to shoulder that burden.

How does the empirical account relate to empiricism? Empiricism asserts that most or all of what T says about observables is true. It entails the relative thesis that T is closer to empirical adequacy than its predecessor. Adherents of the empirical account can accept or reject empiricism. If they reject empiricism, they would be claiming only what they are required to with respect to T. If they accept it, they would be claiming more than they are required to with respect to T. They would not, though, have the burden of refuting the PI. The PI implies that some of the observational consequences of T are false. Hence, the PI is compatible with empiricism, according to which most or all of what T says about observables is true.

Let me turn to the semantic account. It makes the relative claim that T is more truthlike than its forerunner. However, realism makes the absolute claim that T is true. There is a significant difference between the relative and absolute claims. We can infer from the absolute thesis, but not from the relative thesis, that the theoretical entities of T are real, and that most descriptions of the theoretical entities are true, or that they are overall close to the truth.

Advocates of the semantic account can either accept or reject realism without committing a contradiction. If they accept it, they would be claiming more than they are required to with respect to T, and they would needlessly be opening themselves up to the traditional criticisms against realism, such as the PI. By contrast, if they reject realism, they would only be contending that T is more truthlike than its forerunner. This relative claim is compatible with the PI, and thus they would be free from the burden of refuting it.

How does the semantic account relate to empiricism? Advocates of the semantic account can either accept or reject empiricism. If they accept it, they would be claiming more than they are required to with respect to what T says about observables, given that the relative thesis that T is more truthlike than its predecessor does not entail the absolute thesis that most of what T says about observables is true. It is possible that T is more truthlike than its predecessor and yet most of what T says about observables are false. Alternatively, advocates of the semantic account can reject empiricism. If they reject it, they would be claiming that T is more truthlike than its predecessor, but not that T is (approximately) EA. They might add that since T is more truthlike than its predecessor, it is likely that T is closer to the empirical adequacy than its predecessor.

Let me move to the epistemic account. It is substantially different from realism. The epistemic account claims that scientists of T have gained the negative knowledge that the predecessor of T was false and the relative knowledge that T is a better approximation of the truth than its predecessor. By contrast, realism makes the absolute claim that T is true. We can infer from realism, but not from the conjunction of the negative and relative knowledge, that the theoretical entities of T are real.

Defenders of the epistemic account might accept or reject realism. If they accept it, they would have the burden of refuting the standard criticisms against realism, such as the PI. For this reason, Bird (2007) attempts to refute the PI. He writes, "There are many venerable scientific propositions that have never been falsified and which we have no reason to suppose ever will be" (Bird, 2007, p. 101). He also says, "New theories avoid the pitfalls of their falsified predecessors and seek to incorporate their successes" (Bird, 2007, p. 108). Alternatively, defenders of the epistemic account might reject realism. If they reject it, they would only be claiming that scientists of T have the negative and relative knowledge. The negative and relative knowledge is compatible with the PI; consequently, they would not be saddled with the burden of refuting the PI.

Defenders of the epistemic account might accept or reject empiricism. If they accept it, they commit more than they are required to with respect to what T says about observables. After all, the conjunction of the negative and relative knowledge does not entail that T is (approximately) EA. Even if we know that the predecessor of T is false, and that T is more truthlike than its predecessor, T might not be (approximately) EA. By contrast, if defenders of the epistemic account reject empiricism, they would only be maintaining that we have the negative and relative knowledge. Since they claim that we have the relative knowledge, they might add that it is likely that T is closer to empirical adequacy than its predecessor.

What about the noetic account? It is also substantially different from realism. The noetic account makes the relative claim that T explains or predicts more than its predecessor, and thus scientists of T understand more aspects of the world did than their predecessors. By contrast, realism makes the absolute claim that T is true. Moreover, realists believe that T is true. According to Dellsén, however, understanding does not require belief. Therefore, the noetic account is incompatible with realism. It is also incompatible with empiricism because empiricists believe that T is (approximately) EA.

What if other exponents of the noetic account claim that understanding requires belief and justification? They can accept realism without committing a contradiction, given that the relative belief that scientists of T understand more aspects of the world than their predecessors does not preclude the absolute belief that T is true. If they accept realism, however, they would have the burden of refuting the standard objections to realism. Alternatively, they can reject realism, in which case they would only be maintaining that scientists of T understand more aspects of the world than did their predecessors.

These other exponents of the noetic account can accept empiricism without committing a contradiction, given that the relative belief that scientists of T understand more aspects of the world than did their predecessors does not preclude the absolute belief that T is (approximately) EA. If they accept empiricism, however, they commit more than they are required to with respect to what T says about observables. After all, the thesis that scientists of T understand more aspects of the world than their predecessors does not entail that T is (approximately) EA. Alternatively, if they reject empiricism, they would only be maintaining that scientists of T understand more aspects of the world than did their predecessors.

Let me move to the evidential account. It is also substantially different from realism. While the evidential account makes the relative claim that we have more evidence for T than for its predecessor, realism makes the absolute claim that T is true. We can infer from the absolute thesis, but not from the relative thesis, that the theoretical entities of T are real and that most descriptions of the theoretical entities are true or that they are overall close to the truth.

Espousers of the evidential account can either accept or reject realism. They can accept it, given that the relative belief that T is better supported than its forerunner does not preclude the absolute belief that T is true. If they accept it, however, they would have the burden of rebutting the standard criticisms against realism, such as the PI. Alternatively, they can reject it. If they do, they would only be making the modest claim that T is better supported than its forerunner. This modest claim is compatible with the PI. Even if T is fated to be overthrown, espousers of the evidential account can maintain that T is better supported than its forerunner.

The evidential account and empiricism make substantially different claims about what T says about observables. The evidential account makes the relative claim that we have more evidence for what T says about observables than for what its predecessor says about observables. By contrast, empiricism makes the absolute claim that most or all of what it says about observables is true. Espousers of the evidential account can either accept or reject empiricism. If they accept it, they commit more than they are required to with respect to what T says about observables. By contrast, if they reject it, they would only be making the relative claim that T has more evidence than its predecessor.

## 8.9 Summary

Was Semmelweis making scientific progress when he tested the five false hypotheses? According to the problem-solving account, he was not, because none of the five hypotheses was better than its predecessor in terms of the capacity to bring down the high mortality rate. According to the semantic account, he was not, because none of them was closer to the truth than its forerunner. According to the noetic account, he was not, because none of them provided a better understanding of the high mortality rate than its predecessor. According to the epistemic account, by contrast, he was, because he acquired the negative knowledge that a discarded hypothesis was false. According to the evidential account, he was, because the demise of a hypothesis constituted second-order evidence for the next hypothesis.

Why should an account of scientific progress imply that Semmelweis was making scientific progress? Scientists often have failures before attaining a scientific achievement. The more they have failures, the closer they get to a scientific achievement. To get closer to a scientific achievement is to make scientific progress. The same goes for philosophers of science. The more they have philosophical failures, the closer they get to making a philosophical achievement. Just as the demise of logical positivism was a step toward an unproblematic theory of science, so the demise of Semmelweis's false hypothesis was a step toward attaining the true hypothesis.

How do these five accounts relate to realism and empiricism? The problem-solving and noetic accounts are incompatible with realism and empiricism. Therefore, proponents of these two accounts must reject both realism and empiricism. By contrast, the semantic, epistemic, and evidential accounts are compatible with realism and empiricism. Therefore, proponents of these three accounts can accept realism and empiricism without committing a contradiction. However, these accounts do not require realism or empiricism. Therefore, the principle of economy dictates that the proponents not endorse realism or empiricism.

## References

Bird, A. (2007). What is scientific progress? *Noûs, 41*(1), 92–117.

Dellsén, F. (2016). Scientific progress: Knowledge vs. understanding. *Studies in History and Philosophy of Science, 56*, 72–83.

Dellsén, F. (2017). Understanding without justification or belief. *Ratio: An International Journal of Analytic Philosophy, 30*(3), 239–254.

Dellsén, F. (2018). Scientific progress: Four accounts. *Philosophy Compass, 13*(11). https://doi.org/10.1111/phc3.12525

Frost-Arnold, G. (2019). How to be a historically motivated antirealist: The problem of misleading evidence. *Philosophy of Science, 86*(5), 906–917.

Hempel, C. (1966). *Philosophy of natural science*. Prentice-Hall.

Kitcher, P. (1993). *The advancement of science: Science without legend, objectivity without illusions*. Oxford University Press.

Kuhn, T. (1962/1970). *The structure of scientific revolutions*. University of Chicago Press.

Laudan, L. (1977). *Progress and its problems: Towards a theory of scientific growth*. University of California Press.

Laudan, L. (1981). A problem-solving approach to scientific progress. In I. Hacking (Ed.), *Scientific revolutions* (pp. 144–155). Oxford University Press.

Laudan, L. (1984). *Science and values*. University of California Press.

Maxwell, G. (1962). The ontological status of theoretical entities. In H. Feigl & G. Maxwell (Eds.), *Minnesota studies in the philosophy of science III* (pp. 181–192). University of Minnesota Press.

Mizrahi, M. (2013). The argument from underconsideration and relative realism. *International Studies in the Philosophy of Science, 27*(4), 393–407.

Müller, F. (2015). The pessimistic meta-induction: Obsolete through scientific progress? *International Studies in the Philosophy of Science, 29*(4), 393–412.

Niiniluoto, I. (1984). *Is science progressive?* Reidel.

Niiniluoto, I. (2014). Scientific progress as increasing verisimilitude. *Studies in History and Philosophy of Science, 46*, 73–77.

Niiniluoto, I. (2015). Scientific progress. In E. N. Zalta (Ed.), *The Stanford encyclopedia of philosophy*. https://plato.stanford.edu/archives/sum2015/entries/scientific-progress/

Park, S. (2016). Refutations of the two pessimistic inductions. *Philosophia, 44*(3), 835–844.

Park, S. (2017). The uniformity principle vs. the disuniformity principle. *Acta Analytica, 32*(2), 213–222.

Park, S. (2018). In defense of the epistemic imperative. *Axiomathes, 28*(4), 435–446.

Park, S. (2020). Scientific understanding, fictional understanding, and scientific progress. *Journal for General Philosophy of Science, 51*(1), 173–184.

Psillos, S. (1999). *Scientific realism: How science tracks truth*. Routledge.

Ruhmkorff, S. (2013). Global and local pessimistic meta-inductions. *International Studies in the Philosophy of Science, 27*(4), 409–428.

Sankey, H. (2009). Scientific realism and the semantic incommensurability thesis. *Studies in History and Philosophy of Science Part A, 40*(2), 196–202.

Shubin, N., Daeschler, E., & Jenkins, F. (2006). The pectoral fin of *Tiktaalik roseae* and the origin of the tetrapod limb. *Nature, 440*(7085), 764–771.

van der Well, A., van de Ven, G., Maseda, M., Rix, H., Rudnick, G., Grazian, A., Finkelstein, S., Koo, D., Faber, S., Ferguson, H., Koekernoer, A., Grogin, N., & Kocevski, D. (2013). Discovery of a quadruple lens in CANDELS with a record lens red shift Z=1.53. *Astrophysical Journal Letters, 777*.

van Fraassen, B. (1980). *The scientific image*. Oxford University Press.

Wray, B. (2008). The argument from underconsideration as grounds for anti-realism: A defence. *International Studies in the Philosophy of Science, 22*(3), 317–326.

# Chapter 9
# Scientific Realism and Mathematical Realism

**Abstract** It is a fallacious inference that since puzzling sentences result from the attribution of spatial and temporal predicates to mathematical objects, mathematical objects exist in the abstract world. The abstract world is queer, allowing for contradictory states of affairs. The notion of a tricle undermines mathematical realists' idea that the abstract world renders mathematical statements true. Some mathematical realists contend that mathematical objects do not have the properties of being necessary and noncausal, and thus they have the burden of specifying the defining characteristics of mathematical objects. It is to multiply cognitive faculties and entities beyond necessity to say that the cognitive faculty of reason enables us to grasp mathematical objects. There is no empirical evidence for causal platonism, according to which mathematical objects, although abstracta, causally affect human brains. The indispensability argument for mathematical realism has multiple problems. Mathematical inferentialism asserts that a mathematical sentence facilitates deductive inferences between synthetic statements. Mathematical inferentialism is more virtuous than mathematical fictionalism.

**Keywords** Abstract world · Causal platonism · Concrete world · Indispensability argument · Mathematical convenientism · Mathematical inferentialism · Mathematical objects · Tricle

## 9.1 Introduction

Physics, chemistry, biology, and psychology are empirical sciences. Scientists in these fields perform experiments to gather empirical evidence for their hypotheses. Their hypotheses invoke theoretical entities, such as electrons and DNA. By contrast, mathematics is a formal science. Mathematicians do not perform experiments to gather evidence for their statements. Their statements appear to be descriptions of mathematical entities, such as squares, functions, triangles, and numbers. Is mathematics an objective enterprise like empirical science? Can we say that mathematical entities exist like theoretical entities? Mathematical realists (mathematical



S. Park, *Embracing Scientific Realism*, Synthese Library 445,
https://doi.org/10.1007/978-3-030-87813-9_9

platonists) say "Yes," whereas mathematical antirealists (mathematical nominalists) say "No," to these questions. This chapter criticizes mathematical realism and defends mathematical antirealism.

This chapter is organized as follows. In Sect. 9.2, I clarify what it means to say that mathematical entities inhabit the abstract world and that they make mathematical sentences true. In Sect. 9.3, I raise four objections to the suggestion that mathematical objects are abstracta. In Sect. 9.4, I argue that it is problematic to postulate the existence of the cognitive faculty of reason that allegedly specializes in grasping abstracta. In Sect. 9.5, I argue that there is no empirical evidence for causal platonism (Callard, 2007, 2018), according to which mathematical objects causally influence human brains. In Sect. 9.6, I raise three objections to the Quine-Putnam indispensability argument. In Sect. 9.7, I introduce mathematical inferentialism (Park, 2017a) as an alternative to mathematical realism and fictionalism. The main thesis of this chapter is that scientific realists should choose mathematical inferentialism over mathematical realism and fictionalism.

## 9.2 Mathematical Realism

Let me distinguish between concrete and abstract objects. Concrete objects are spatial, temporal, causal, and mutable. Consider, for example, two billiard balls, $b_1$ and $b_2$, which are located in space and time. At $t_1$, $b_1$ is in a specific place, and $b_2$ is in another place. In this sense, they are spatial and temporal. At $t_2$, $b_1$ moves to $b_2$, which is at rest, and bumps into $b_2$. As a result, $b_1$ changes its direction, and $b_2$ starts to move. In this sense, they are causal and mutable. Concrete objects exist in the concrete world.

By contrast, abstract objects exist in the abstract world. Abstracta are nonspatial, immutable, nontemporal, and noncausal (Balaguer, 2018; Linnebo, 2018). Consider, for example, two mathematical objects, a triangle and a square. It is a conceptual mistake to claim that a triangle exists in a specific place while a square exists in another place. It is also a category mistake to contend that they existed yesterday. In this sense, they are nonspatial and nontemporal. In addition, they cannot bump into each other, nor can they collide with concrete objects. In this sense, they are noncausal and immutable.

Let me now differentiate between concrete and abstract statements. Consider, for example, the concrete statement that Jupiter revolves around the sun. This statement is true whether we think that it is true or not. It is not our thinking but the concrete world that makes it true. In this sense, astronomy is an objective enterprise. Consider, for example, the abstract statement that $-1x-1=1$. Mathematical realists suggest that this mathematical statement is true whether we think that it is true or not. What makes it true? To answer this question, they hypothesize that the abstract world exists. It is true that $-1x-1=1$ because the statement corresponds to a mathematical fact in the abstract world, just as it is true that Jupiter revolves around the sun because the statement corresponds to a concrete fact in the concrete world. Thus, mathematics is an enterprise that is just as objective as empirical science.

How would entity realists respond to mathematical realism? Recall that according to entity realism, if we can manipulate certain theoretical entities, "then they are real" (Hacking, 1983, p. 23). We can say, for example, that electrons and protons are real because we can manipulate them. Now, it is in principle impossible to manipulate mathematical objects, given that they are abstracta. Therefore, entity realists would flatly reject mathematical realism. As we have seen in Sect. 5.2 of Chap. 5, however, entity realism has its own problems. In the subsequent sections of this chapter, I criticize mathematical realism without relying on entity realism.

## 9.3 The Abstract World

### *9.3.1 The Argument from Language*

On what grounds do mathematical realists contend that mathematical objects, such as the cosine function, are nonspatial and nontemporal? They observe that "It makes no sense to ask where the cosine function was last Tuesday" (Rosen, 2018). In other words, meaningless sentences result from the attributions of spatial and temporal predicates to mathematical objects. Therefore, mathematical objects are nonspatial and nontemporal. This argument for mathematical realism might be called "the argument from language."

Let me use an analogy to show that the argument from language is incorrect. Suppose that an astronomer is interested in the moon. Instead of investigating the moon, however, the astronomer studies your finger, which points at the moon, and then argues that since there is blood in your finger, there must also be blood in the moon. This inference is absurd. Now, mathematical realists are interested in mathematical objects. Instead of investigating mathematical objects, however, they investigate mathematical sentences and then argue that since puzzling sentences arise from the attribution of temporal and spatial predicates to mathematical objects, mathematical objects must be nonspatial and nontemporal. This inference is as unacceptable as the previous inference from the finger to the moon.

If you are interested in mathematical objects, you should investigate mathematical objects, not mathematical sentences. You should present evidence to show that mathematical objects are nonspatial and nontemporal, instead of saying that puzzling sentences result from the attribution of spatial and temporal predicates to mathematical objects. The fact that it makes no sense to say that the cosine function was here last Monday is a fact about a mathematical sentence, not about the cosine function. What is true of a sentence may not be true of a state of affairs referenced by the sentence.

### 9.3.2 The Queer World

The abstract world is queer (Park, 2017b, pp. 99–101). Euclidean geometry, Lobachevskian geometry, and Riemannian geometry are all incompatible with one another. Consider the three angles of a triangle. According to Euclidean, Lobachevskian, and Riemannian geometries, their sum is 180°, smaller than 180°, and larger than 180°, respectively. Mathematical realists tell us that these three mathematical statements are true because they correspond to three mathematical facts in the abstract world. In my view, however, these three mathematical states of affairs cannot coexist in the abstract world, and to say that they do implies that the abstract world is queer.

Mathematical realists might reply that there are three abstract worlds: the Euclidean world, the Lobachevskian world, and the Riemannian world. Each of these worlds makes one of the three geometries true; it is wrong to think that one abstract world renders all three geometries true. In my view, however, this reply multiplies the worlds without any independent reason, and thus it opens to the door for empirical scientists to make a similar move to avoid the refutation of their preferred scientific theory. For example, ether theorists might argue that the ether theory is true since there is an ether world that renders it true (Park, 2017b, pp. 100–101).

Let me add here that to postulate the existence of the Euclidean, Lobachevskian, and Riemannian worlds raises a question about how we should individuate abstract worlds. One might suggest that there are as many Euclidean abstract worlds as there are Euclidean statements, and each Euclidean statement is rendered true by one Euclidean abstract world. The same suggestion could be made about Lobachevskian geometry and Riemannian geometry. If the existence of such abstract worlds is dubious, the existence of the three abstract worlds for each kind of geometry is also dubious. So is the existence of the abstract world to which all true mathematical statements allegedly correspond.

A reviewer advances an interesting critique of my claim that it is absurd for ether theorists to say that the ether theory is true since there is an ether world that renders it true. On a neo-Kantian interpretation of Thomas Kuhn (1962/1970), it might not be absurd for ether theorists to say so. In the phenomenal world of the ether theory, certain claims about the ether are true, whereas in the phenomenal world of non-ether theories, those claims are false.

Regarding the reviewer's comment, selective realists would say that we do not need to invoke the neo-Kantian notion of the phenomenal world to argue that some theoretical claims of the ether theory are true, while other theoretical claims of it are false. As far as selective realists are concerned, there is only one world, and it renders some theoretical claims of the ether theory true and other theoretical claims of it false. The former claims are called "working posits," and the latter claims are called "idle posits," as we saw in Subsect. 5.6.1 of Chap. 5. Hence, it is to multiply worlds beyond necessity to postulate phenomenal worlds and the noumenal world.

I previously stated that ether theorists might argue absurdly that the ether theory is true since there is an ether world that renders it true. I admit that my language was not precise enough. I now say that ether theorists might argue absurdly that all theoretical claims of the ether theory are true since there is an ether world that renders them true. In my view, even neo-Kantians would reject this argument. After all, the ether theory conflicts with the outcome of the Michelson-Morley experiment, and hence it is false of its phenomenal world.

### *9.3.3 Tricle*

According to mathematical realism, mathematical objects are abstracta, and mathematical facts, which are part of the abstract world, make mathematical statements true. Mathematical realists invoke the abstract world to establish that mathematics is as objective an enterprise as empirical science. This subsection exposes a problem with this strategy to secure the objectivity of mathematical truths.

Imagine that an object called a "tricle" (triangle and circle) exists. It changes its shape every second from a circle to a triangle and from a triangle back to a circle (Park, 2017b, pp. 102–103). A tricle is noncausal and nonspatial, for both a circle and a triangle are noncausal and nonspatial. However, it is temporal and mutable, for it changes its shape as time passes. The following statement about a tricle is true:

(T) The sum of the interior angles of a tricle is sometimes 180° and at other times 360°.

What makes (T) true? It is wrong to answer that the abstract world renders it true because the abstract world is a nontemporal world. It is also problematic to say that (T) is true by virtue of the definition of a "tricle." After all, if mathematical realists said so, they would also have to say that the following statement is true by virtue of the definition of "square":

(S) The sum of the four angles of a square is 360°.

If (S) were analytic, all other mathematical statements would also be analytic. If all mathematical statements were analytic, there would be no need to postulate the existence of mathematical objects, and mathematical realism would collapse.

Mathematical realists might suggest that there is a third world, which is neither the abstract world nor the concrete world. The denizens of the third world are nonspatial, temporal, noncausal, and changing. A tricle inhabits this third world. A state of affairs in this third world makes (T) true. Thus, (T) is not analytic but synthetic.

However, it is ad hoc to postulate the existence of a third world to handle (T). An independent reason is required for postulating the existence of a third world (Park, 2019, p. 252). After all, some theorists might imagine an object that is nontemporal and nonspatial but causal. If they did, mathematical realists would have to postulate the existence of a fourth world that is purportedly inhabited by nontemporal and nonspatial but causal objects. However, such a move is open to the charge that mathematical realists multiply worlds beyond necessity.

The problem of multiplying worlds beyond necessity looms large when we consider the following false statement:

(N) A tricle does not change its shape.

What makes (N) false? Mathematical realists might suggest that it is false because it does not correspond to the third world. However, why not say that it is false because it does not correspond to the abstract world? Similarly, consider the following mathematical statement:

(M) $-1\text{x}-1 = 1$

Mathematical realists say that (M) is true because it corresponds to the abstract world. However, why not say that it is false because it does not correspond to the third world?

Mathematical realists owe us answers to these questions because they invoke the abstract world to establish that the truth-value of a mathematical statement is as objective as that of a concrete sentence. If it were legitimate to invoke the abstract world, it should also be legitimate to invoke a third world, a fourth world, and so on, and it becomes unclear which world we should choose to determine whether a statement is true or false. I suppose that mathematical realists would reject the absurd suggestion that a statement changes its truth-value, depending on which world we choose as the truth-maker or the falsity-maker. After all, to accept this suggestion goes against their beliefs that mathematics is an objective enterprise like empirical science. If they reject the absurd suggestion, however, they have the burden of specifying the criterion for choosing one world over another as the truth-maker of a statement.

### *9.3.4 The Essential Properties of Mathematical Objects*

Mathematical realists initially attributed five properties to mathematical objects, viz., being nontemporal, nonspatial, noncausal, immutable, and necessary. However, they have been jettisoning some of the five properties in the recent history of philosophy of mathematics. Indispensabilists like Willard Quine contend that mathematical objects are, though abstract, not necessary. They maintain that mathematical statements are subject to empirical refutation. Another mathematical realist, Benjamin Callard (2007, 2018), claims that mathematical objects, though abstract, are causally efficacious on human brains, as we will see in Sect. 9.5 below.

Based on this pattern by mathematical realists, one might predict that mathematical realists will eliminate more properties from the list of the five properties. This prediction might turn out to be false. Even if it does, the fact that mathematical realists have jettisoned the properties of being necessary and noncausal raises an interesting issue. What are the defining characteristics of a mathematical object? Which of the five properties above are essential and which are accidental? Imagine that an object does not have the property of being nonspatial, but it does have the remaining four properties. Is it a mathematical object? Mathematical realists owe us answers to these questions.

## 9.4 The Epistemological Objection

It is clear how we form knowledge about concrete objects, such as snow and the moon. They causally affect our brains. By contrast, abstract objects, such as propositions and mathematical objects, are causally inert, and thus they cannot causally affect human brains. Therefore, mathematical realists are saddled with the burden of giving an account of how we form knowledge about mathematical entities (Benacerraf, 1973).

Rationalists tell us that mathematical knowledge is a priori, which means that the cognitive faculty of reason is the source of the justification for a mathematical belief. The mind somehow grasps a mathematical object in the abstract world, although the mathematical object does not causally affect the human brain. We use mental eyes, so to speak, to see that $-1x - 1 = 1$, just as we use physical eyes to see that snow is white.

A standard objection to the rationalist account of mathematical knowledge is that it is mysterious how the mind grasps a mathematical object. A mathematical object is causally inefficacious, which implies that it cannot causally affect the mind. It is not clear how the mind grasps a mathematical object without being causally affected by it. It appears to be a contradiction to say that the mind grasps a mathematical object, but it is not causally affected by it.

Let me raise a related objection to the rationalist account of mathematical knowledge. How much matter is there in the universe? Rationalists claim that the mind can grasp infinitely many numbers in the abstract world. If the mind has such an amazing ability, it should also have the ability to grasp the amount of matter in the universe. Rationalists would answer that the mind specializes in grasping abstract objects, but not concrete objects, so we do not know how much matter exists in the concrete world, although we do know how many natural numbers there are in the abstract world. In response, I propose an alternative answer. Just as we do not know how much matter exists in the concrete world, so we do not know how many numbers there are in the abstract world. We do not have the mathematical knowledge because there is no such thing as the abstract world. In general, we cannot have knowledge about what does not exist.

Imagine that some claim that ghosts exist, that their beliefs about ghosts are justified, and that the justification for their beliefs comes from a cognitive faculty that they dub "the g-faculty." They also claim that unicorns exist, that their beliefs about unicorns are justified, and that the justification for their beliefs comes from the cognitive faculty that they call "the u-faculty." You would immediately think that they multiply cognitive faculties as well as entities beyond necessity. If you think so, however, you should also think that rationalists multiply cognitive faculties and entities beyond necessity. After all, rationalists' reasoning is no different from that of those who claim that ghosts and unicorns exist.

On what grounds do we postulate the existence of the faculty of vision? Suppose that we seem to see an apple on a tree and that as a result, we believe that there is an apple on a tree. This visual belief can be confirmed by other faculties: touch, smell, taste, and hearing. Moreover, we can establish a causal relationship between the apple and human brains. However, similar actions cannot be taken about the faculty of reason that allegedly specializes in grasping the objects in the abstract world. Suppose that we believe that $-1x - 1 = 1$. This mathematical belief cannot be confirmed by other cognitive faculties, nor can we establish the causality between mathematical entities and human brains. The onus is on mathematical realists to show that just as it is legitimate to postulate the existence of vision, so it is legitimate to postulate the existence of the faculty of reason that allegedly specializes in grasping mathematical objects.

## 9.5 Causal Platonism

In order to get around Paul Benacerraf's (1973) epistemological objection to mathematical realism above, Callard (2018) defends an original position that he calls "causal platonism." It holds that "Abstract mathematical objects efficiently cause changes in the brain" (Callard, 2018, p. 11). Put differently, "Numbers causally affect human brains" (Callard, 2018, p. 13). Causal platonism differs from the traditional variant of mathematical realism, according to which mathematical objects are noncausal. Causal platonism does not merely claim that mathematical objects *can* causally affect human brains. Instead, it claims that they *do* causally affect human brains. In other words, according to causal platonism, it is not merely metaphysically possible, but actually the case, that mathematical objects causally affect human brains.

An immediate objection to causal platonism is that it is an empirical claim that mathematical objects make changes in the brain, so there should be empirical evidence for causal platonism. However, we cannot gather empirical evidence for causal platonism (Park, 2019, p. 250). Suppose that certain neurons in your brain are activated and that as a result, you form the mathematical knowledge that $-1x - 1 = 1$. How should neuroscientists go about investigating the mathematical cause of this brain state? They have no idea where and when to look, given that the mathematical objects that purportedly caused the brain state are nonspatial and nontemporal. Therefore, there can be no empirical evidence establishing the causal relationship between human brains and mathematical objects.

Surprisingly, Callard claims that causal platonism is justified not a posteriori but rather a priori. He says, "We can know, without empirical confirmation, that causal platonism is true" (Callard, 2018, p. 8). We can know that mathematical objects are abstracta not by observing the concrete world, but by grasping mathematical objects, i.e., "not by doing empirical fieldwork, but by grasping, and making inferences from, the content of mathematical judgments" (Callard, 2018, p. 7). We can thus know that mathematical objects cause mathematical knowledge not by observing the concrete world but by observing mathematical knowledge. In other words, causal platonism is derived "from reflection on the metaphysical character of knowledge, and not from empirical investigations of material objects or the laws of nature" (Callard, 2018, pp. 7–8). In sum, we know a priori, not a posteriori, that mathematical objects cause mathematical knowledge (Callard, 2018, pp. 6–7).

How do I respond to this original argument for causal platonism? Imagine that astronomers claim that the moon affects the motion of terrestrial objects. Instead of investigating the moon and terrestrial objects, however, they investigate the finger pointing to the moon to justify their claim. We would point out that their claim can only be established by investigating the moon and terrestrial objects. Now, causal platonists claim that mathematical objects causally affect human brains. Instead of investigating mathematical objects and human brains, however, they investigate mathematical knowledge to justify their claim. We should point out that their claim can only be established by investigating mathematical objects and human brains.

Moreover, Callard has slid from brain states to mathematical knowledge. He originally defines causal platonism as the view that mathematical objects produce *brain states*. As cited in the first paragraph of this section, he states, "Abstract mathematical objects efficiently cause changes in the brain" (Callard, 2018, p. 11), and that "Numbers causally affect human brains" (Callard, 2018, p. 13). However, as cited in the third paragraph of this section, he now states that mathematical objects produce *mathematical knowledge*. He replaces "brain states" with "mathematical knowledge" in the context of responding to my objection (Park, 2019, p. 250) that there should be empirical evidence for causal platonism.

Why do I fuss about this slide? First, there is a vast difference between the thesis that mathematical objects produce brain states and the thesis that they produce mathematical knowledge. Brain states are physical states, whereas mathematical knowledge is a mental state. I should observe my brain to determine whether I am in a certain brain state. A scientific instrument, such as an MRI scanner, is needed for me to observe my own brain state. By contrast, I only need to introspect my mind to determine whether I have certain mathematical knowledge. It follows that empirical evidence is required to show that I am in a certain brain state, whereas no empirical evidence is required to show that I have certain mathematical knowledge.

Second, it is to beg the question against mathematical antirealists to say that we know a priori that causal platonism is true. It is under dispute between mathematical realists and antirealists whether we have mathematical knowledge that is putatively about mathematical objects in the abstract world. In the attempt to prove that mathematical objects exist, however, Callard assumes that we have such knowledge, which implies that mathematical entities inhabit the abstract world.

Callard's argument for causal platonism is similar to the following circular argument. Dragon theorists believe that dragons exist. They argue that dragons cause their beliefs about them, that their beliefs about them are justified, and that the justification for their beliefs comes from the d-faculty. Moreover, they argue that this story of how their beliefs about dragons are justified is true and that the justification for the true story comes from the d-faculty. Both Callard's argument and dragon theorists' argument are circular, and they both multiply cognitive faculties and entities beyond necessity.

Moreover, suppose that it is an a priori thesis, as Callard argues, that mathematical objects cause mathematical knowledge. It does not follow that the a priori thesis is true. It is one thing to say that a thesis is a priori; it is another to say that it is a priori true. For example, it is an a priori statement that a bachelor is married; however, it is a false a priori statement. Similarly, it might be a false a priori thesis that mathematical objects cause mathematical knowledge. What needs to be established is not that causal platonism is a priori, but that it is a priori true.

In summary, causal platonism is an a posteriori thesis, given that it is a thesis about human brains. Hence, it needs to be justified a posteriori. However, there cannot be empirical evidence for it, given that we have no idea when and where to look to establish the causal relationship between human brains and mathematical objects. Callard gives an a priori justification for causal platonism. However, he slides from brain states to mathematical knowledge, thereby begging the question against mathematical antirealists and running a circular argument.

## 9.6 The Indispensability Argument

### *9.6.1 Content*

The previous sections show, I hope, that the a priori account of how we form mathematical knowledge is not convincing. Fortunately, some eminent philosophers of mathematics have given an a posteriori account of how we form mathematical knowledge. They include Quine (1948, 1980, 1992), Hilary Putnam (1971), Michael Resnik (1997), and Mark Colyvan (2001). They defend the famous argument for mathematical realism called "the Quine-Putnam indispensability argument." This subsection explicates it, and the subsequent subsections criticize it.

What does the indispensability argument say? It can be roughly summarized in the following standard form:

Mathematical sentences are indispensable ingredients of our best theories.
All constituents of our best theories, including mathematical constituents, are confirmed by observational evidence.
∴We can justifiably believe that mathematical entities exist.

Scientific realists believe that theoretical entities exist on the grounds that theoretical constituents of our best theories are confirmed by observational evidence. Therefore, they should also believe that mathematical entities exist. It would be intellectually dishonest for them not to believe that mathematical entities exist (Putnam, 1979, p. 347; Quine, 1980, p. 45; Colyvan, 2006, pp. 226–227).

The second premise, which goes by "confirmational holism," is built upon the view that mathematical statements are synthetic. If they were analytic, it would be false that the mathematical constituents as well as the theoretical constituents of our best theories are confirmed by observational evidence. Confirmational holism goes hand in hand with Quine's (1951) obliteration of the distinction between analytic and synthetic statements and is contrary to logical positivists' view that mathematical statements are analytic. A synthetic statement can be confirmed or disconfirmed by observational evidence, while an analytic statement cannot.

The indispensability argument, if correct, shows that mathematical objects exist; it does not show, however, that they exist as abstracta. It is one thing to say that mathematical entities exist; it is quite another to say that they are nonspatial and nontemporal. After all, there are other ways that mathematical objects may exist. For example, they might be spatial but nontemporal; alternatively, they might be nonspatial but temporal, or they might be both spatial and temporal. The indispensability argument is reticent about what kind of objects they are. Therefore, when indispensabilists claim that mathematical objects are abstracta, they are claiming more than what their argument licenses them to do.

The indispensability argument has a virtue: it provides an admirable answer to Benacerraf's objection to mathematical realism, viz., we acquire mathematical knowledge not by being causally affected by mathematical entities but rather by observing the concrete world. It follows that mathematical realists do not need to give an account of how abstract entities causally affect human brains.

### 9.6.2 *Standard Objections*

The first premise of the indispensability argument says that mathematical sentences are indispensable constituents of our best theories. This premise is disputed by Hartry Field (1980), who attempts to erase the mention of mathematical objects from Isaac Newton's theory of gravity. Some philosophers defend, while other philosophers criticize, Field's project. At the moment, Field's project "remains controversial" (Balaguer, 2018). I do not pursue this line of attack on the indispensability argument in this book.

The indispensability argument clashes with the pessimistic induction (PI). If the PI is cogent, our best theories will turn out be false. The demise of our best theories means that just as theoretical entities, such as caloric and phlogiston, turned out to be nonexistent, so mathematical entities, such as the cosine function and numbers, will turn out to be nonexistent. Therefore, indispensabilists have the burden of refuting the PI, just as do scientific realists. Relatedly, scientific realists cannot invoke the PI to refute the indispensability argument because the PI refutes not only the indispensability argument but also scientific realism.

The second premise of the indispensability argument says that all components of a scientific theory are confirmed by empirical evidence. This premise is challenged by Penelope Maddy and Jacob Busch. Maddy (1992) notes that some scientific theories invoke ideal objects, such as spherical molecules and frictionless slopes, and we do not believe that ideal objects exist. Busch (2012) observes that according to selective realism, the working constituents of a scientific theory are confirmed by observation while its idle constituents are not.

Maddy's argument rebuts confirmational holism. However, her argument has a certain limitation, viz., it is silent about whether mathematical objects are in the same boat as real objects or as ideal objects, i.e., whether mathematical objects are the kind of objects whose existence is confirmed by observations or the kind of objects whose existence is not confirmed by observations. Indispensabilists might maintain that mathematical objects are in the same boat as real objects, but not as ideal objects. In short, it is one thing to say that not all objects postulated by our best theories are real; it is entirely another to say that mathematical entities are not real. Mathematical antirealists have the burden of arguing that mathematical objects are in the same boat as ideal objects and that just as we do not believe that ideal objects are real, so we should not believe that mathematical entities are real.

The same goes for Busch's argument. His argument refutes confirmational holism; however, it has a similar limitation. It is reticent about whether mathematical assumptions are in the same boat as working assumptions or as idle assumptions. Indispensabilists might maintain that mathematical assumptions are in the same boat as working assumptions, but not as idle assumptions. In short, it is one thing to say that not all constituents of our best theories are confirmed by observations; it is entirely another to say that mathematical constituents are not confirmed by observations. Mathematical antirealists have the burden of arguing that mathematical assumptions are in the same boat as idle assumptions and that just as we should not believe that idle assumptions are true, so we should not believe that mathematical assumptions are true. I doubt that mathematical antirealists can establish that mathematical assumptions are in the same boat as idle assumptions. After all, mathematical assumptions have survived many scientific revolutions in the history of science, just as have working assumptions.

Moreover, Alan Baker (2005, 2009, 2012, 2016) has developed what he calls "the enhanced indispensability argument," which claims that a mathematical statement can be directly confirmed by empirical evidence. For example, the nymphal stage of periodic cicadas is either 13 or 17 because the nymphal stage of these lengths minimizes their interactions with predators. If the nymphal stage were neither 13 nor 17, the cicadas would be more likely to be eaten by predators. It is debated whether Baker's example is a genuine mathematical explanation of physical phenomena. In any event, Baker's argument for mathematical realism does not rely on confirmational holism, and thus it presents a new challenge to mathematical antirealists, such as Maddy and Busch.

### 9.6.3 The Two Distinctions

Quine (1951) attacks the distinction between analytic and synthetic statements, and he (1948, 1980, 1992) defends the indispensability argument, distinguishing between concrete and abstract sentences. In this subsection, I explore whether it is justifiable for indispensabilists to reject the distinction between analytic and synthetic statements yet to accept the distinction between concrete and abstract statements.

In contrast to Quine, logical positivists suggest that mathematical statements are analytic. A standard objection to this suggestion is that it is mysterious how mathematical statements are useful in empirical science. Analytic statements are empty of content about the world, i.e., they do not make any claims about the world. Empirical science is an enterprise that investigates the world. Therefore, logical positivists owe us an explanation of how statements that do not even purport to describe the world are useful in an enterprise that describes the world.

In my view, however, a similar objection can be raised against the suggestion that mathematical statements are abstract, i.e., that they are descriptions of abstract objects. Mathematical statements are empty of content about the concrete world, although they might have content about the abstract world. Empirical science is an enterprise that investigates not the abstract world but the concrete world. It is mysterious how statements that do not even purport to describe the concrete world can be useful in an enterprise that describes the concrete world (Park, 2017a, pp. 77–78).

Indispensabilists might reply that mathematical sentences are useful in empirical science, although they do not make any claims about the concrete world, because mathematical sentences are indispensable parts of a scientific theory that is, as a whole, useful in empirical science. However, this reply would evoke an equally defensive response from logical positivists, who believe that mathematical statements are analytic. They would suggest that mathematical statements are useful in empirical science, although they do not make any claims about the world, because they are indispensable parts of a scientific theory that is, as a whole, useful in empirical science.

Indispensabilists might object that analytic statements, although they might be indispensable components of a scientific theory, cannot be confirmed by observations because they do not make any claims about the concrete world. How can sentences that are empty of information about the concrete world be confirmed by observations? In my view, however, this objection would elicit an equally critical response from logical positivists. They would argue that mathematical sentences, although they are indispensable elements of a scientific theory, cannot be confirmed by observations because they do not make any claims about the concrete world. How can sentences that are empty of information about the concrete world be confirmed by observations?

Indispensabilists might now suggest that the distinction between concrete and abstract statements is tenable whereas the distinction between analytic and synthetic statements is not, so logical positivists cannot say that mathematical statements are analytic. In my view, however, this suggestion calls into question the distinction between concrete and abstract sentences. What is the criterion for distinguishing between concrete and abstract sentences? Indispensabilists might answer that concrete sentences are those that ordinary speakers intend to use to describe the concrete world, while abstract sentences are those that ordinary speakers intend to use to describe the abstract world. A problem with this suggestion is that ordinary language users have no idea what the concrete and abstract worlds are.

Indispensabilists might now suggest that concrete sentences are those that actually describe the concrete world, while abstract sentences are those that actually describe the abstract world. In my view, however, this suggestion begs the question against mathematical antirealists, who deny the existence of the abstract world. It is under dispute between mathematical realists and antirealists whether the abstract world exists. Therefore, mathematical realists cannot invoke the abstract world to distinguish between concrete and abstract sentences. As far as mathematical antirealists are concerned, the abstract world does not exist, and thus the distinction between concrete and abstract sentences is untenable.

Moreover, there are tricky sentences: "The concrete world is inhabited by numbers," "Numbers inhabit the concrete world," and "Numbers and cats exist in the concrete world." Are these sentences abstract or concrete? Mathematical realists might answer that they are about numbers, and thus they are abstract sentences. However, mathematical antirealists have an opposing answer: the sentences are about the concrete world and cats, and thus they are concrete sentences, and they are false. It is not clear whether this dispute can be resolved. It sounds bizarre to say that the tricky sentences are half-concrete or half-abstract sentences or that they are both concrete and abstract sentences.

What can we conclude from the difficulty to distinguish between abstract and concrete sentences? Mathematical antirealists could advance an argument similar to that of Quine (1951). Quine argues that the distinction between analytic and synthetic statements is untenable, so no sentence is analytic, and sentences that we ordinarily take to be analytic are, in fact, synthetic. Mathematical antirealists could similarly argue that the distinction between abstract and concrete sentences is unsustainable, so no sentence is abstract, and sentences that we ordinarily take to be abstract are, in fact, concrete. Since no sentence is abstract, mathematical realism cannot get off the ground.

To summarize, indispensabilists have the burden of explaining how mathematical sentences, which do not make any claims about the concrete world, can nevertheless be useful in empirical science. Whatever account they might give, logical positivists would make use of it to defend the opposing position that mathematical statements are analytic. Indispensabilists might argue that the distinction between analytic and synthetic statements is untenable, so we cannot say that mathematical sentences are analytic. In response, however, mathematical antirealists can argue that the distinction between concrete and abstract statements is untenable, so it is problematic to say that mathematical sentences are abstract.

### 9.6.4 Interactive Theoretical Entities

According to indispensabilists, mathematical entities enjoy the same epistemic status as theoretical entities, and thus it is illegitimate to affirm that theoretical entities exist while denying that mathematical entities exist. I dispute this claim in this subsection. This subsection demonstrates that some theoretical entities enjoy a higher epistemic status than mathematical entities, so it is not intellectually dishonest to affirm the existence of those theoretical entities but to deny the existence of mathematical entities.

We can divide theoretical entities into two groups: interactive and non-interactive theoretical entities (Park, 2018, p. 12). Interactive theoretical entities are theoretical entities whose interactions are invoked to explain new phenomena. DNA and ultraviolet rays are interactive theoretical entities in this sense. Their existence was confirmed independently of each other, and the interaction between them explains why some skin cancers occur (Trefil & Hazen, 1998, pp. 155–156). Compare them with phlogiston, a non-interactive theoretical entity. Scientists did not invoke the interaction between phlogiston and another theoretical entity to explain other phenomena. The same holds for other past theoretical entities, such as the celestial sphere and caloric. Therefore, we have more evidence for DNA and ultraviolet rays than for phlogiston, caloric, or the celestial sphere.

The indispensability argument, even if correct, shows at best that mathematical entities are as highly confirmed as non-interactive theoretical entities, such as phlogiston and caloric. The interaction between two mathematical entities cannot be invoked to explain new phenomena because mathematical entities cannot, by definition, interact with each other and because they cannot, by definition, interact with concrete entities. Therefore, the existence of mathematical entities cannot, by definition, be as highly confirmed as the existence of interactive theoretical entities (Park, 2018, p. 14).

It is not a double standard to believe that interactive theoretical entities exist, but to disbelieve that non-interactive theoretical entities exist. It follows that it is not a double standard to believe that interactive theoretical entities exist, but to disbelieve that mathematical objects exist (Park, 2018, p. 15). In addition, it is a double standard to believe that mathematical entities exist, but to disbelieve that non-interactive theoretical entities, such as the celestial sphere, phlogiston, and caloric, exist. If mathematical realists disbelieve that past theoretical entities exist, they have the burden of pointing out the relevant differences between past theoretical entities and mathematical entities.

Selective realists claim that "phlogiston" *approximately* refers to oxygen, given that some assumptions of the phlogiston theory can be found in the oxygen theory (Psillos, 1999, p. 294). On this account, reference is not an all-or-nothing affair. It is not clear to me whether mathematical realists can make a similar move. It does not sound reasonable to say that some assumptions about a square are true, while other assumptions about it are false, so "square" approximately refers to squares.

### 9.6.5 *Mathematical Convenientism*

The indispensability argument implies that mathematical statements, such as "2+2=4," are subject to empirical refutation. However, "no proposition of pure mathematics has been *falsified*" (Parsons, 1983, p. 196). Why have no purely mathematical statements been thrown out, while many concrete statements, such as "Light travels through the medium of ether," have been thrown out in the history of science? Indispensabilists owe us an answer to this question because they claim that mathematical statements are synthetic, just as are concrete statements.

How do indispensabilists reply to Charles Parsons's (1983) critique? Quine replies that we can falsify mathematical statements, but we do not because "mathematics infiltrates all branches of our system of the world, and its disruption would reverberate intolerably" (Quine, 1992, p. 15). The idea is that negating a mathematical statement and negating a concrete statement have repercussions of different degrees on our system of beliefs. If we negate a concrete statement, only a few beliefs need to be revised. However, if we negate a mathematical statement, many of our beliefs need to be revised. In a similar vein to Quine, Resnik and Colyvan argue, respectively, that falsifying a mathematical statement "is likely to send reverberations into currently quiescent areas of science" (Resnik, 1997, pp. 125–126) and that "it's not that mathematical theories are unfalsifiable in any absolute sense; it's just that whenever they are part of a falsified package, the mathematical portion is unusually salvaged for reasons of minimum mutilation to the web of belief" (Colyvan, 2001, p. 126).

These three philosophers' position is named as "mathematical convenientism" (Park, 2016, p. 118). It says that we falsify concrete beliefs but not mathematical beliefs because that is more convenient for us. Is mathematical convenientism acceptable? It commits a consequential fallacy (Park, 2016, pp. 118–119). A consequential fallacy arises when we take a statement to be true on the grounds that it brings about happiness or when we take a statement to be false on the grounds that it brings about unhappiness. For example, one might take evolutionary theory to be false on the grounds that it is degrading to think that humans and nonhuman organisms were descended from a common ancestor.

Mathematical convenientism asserts that we choose a belief system over another belief system not because the former is more likely to be true than the latter but rather because it is more convenient to do so. It implies that there is no epistemic basis for choosing one belief system over another, which in turn implies that there is no epistemic basis for the elements of our belief system, including our observational beliefs, scientific theories, and mathematical beliefs (Park, 2016, pp. 119–120). We posit, for example, that snow is white not because we have good evidence for the belief but rather because it is a part of our belief system, which we have chosen over other belief systems for the sake of convenience. The same holds for theoretical beliefs and mathematical beliefs. Therefore, mathematical convenientism ultimately destroys mathematical realism.

The preceding discussion shows that it is self-defeating for indispensabilists to invoke mathematical convenientism to diffuse Parsons's objection above, i.e., to give an account of why the history of science witnessed no refutation of a purely mathematical statement. Hence, indispensabilists need to seek an alternative account that does not invoke mathematical convenientism. They cannot propose that no purely mathematical statement has been refuted because purely mathematical statements are analytic. Such an account is not available to them; it is available only to mathematical antirealists.

## 9.7 Mathematical Inferentialism

In the previous sections, I argued that the abstract world is queer. I also criticized rationalists' account, causal platonists' account, and indispensabilists' account of how we form mathematical knowledge. Readers may wonder what my alternative account is. In this section, I enunciate my alternative to mathematical realism, which I call "mathematical inferentialism" (Park, 2017a, pp. 71–73).

Mathematical inferentialism consists of the following five theses. The first is the logical thesis that the conjunction of a mathematical sentence and some synthetic statements entails other synthetic statements. For example, consider the following mathematical sentence and synthetic statements:

(m) 10+10=20
($s_1$) He gave me $10.
($s_2$) She gave me $10.
($s_3$) So they gave me $20 in total.

In conjunction with ($s_1$) and ($s_2$), (m) entails ($s_3$). (M) all by itself does not entail ($s_3$), but it does, once it is supplied with ($s_1$) and ($s_2$). This logical relationship between a mathematical sentence and synthetic statements is similar to the logical relationship between a scientific theory and observational statements. A scientific theory alone does not entail any observational consequences, but it does, once it is combined with auxiliaries: auxiliary assumptions and statements of initial conditions (Duhem, 1905/1954).

The second is the psychological thesis that a mathematical sentence facilitates deductive inferences between synthetic statements (Park, 2017a, p. 71). For example, (m) facilitates the deductive inference from ($s_1$) and ($s_2$) to ($s_3$). This psychological function of a mathematical sentence is similar to that of a scientific theory. As instrumentalists observe, a scientific theory facilitates deductive inferences from some observational statements to other observational statements. Just as this psychological function of a scientific theory is based upon the logical relationship between a scientific theory and observational statements, so the psychological function of a mathematical sentence is based upon the logical relationship between a mathematical sentence and synthetic statements.

The third is the semantic thesis that a mathematical sentence is true when it is *synthetically adequate*. To say so means that a true mathematical sentence, when conjoined with true synthetic statements, entails only other true synthetic statements. To use the example above, it is synthetically adequate that 10 + 10 = 20 because when conjoined with some true synthetic statements, such as ($s_1$) and ($s_2$), it issues only other true synthetic statements, such as ($s_3$). Accordingly, it is true that 10 + 10 = 20. By contrast, it is synthetically inadequate that 10 + 10 = 30 because when supplied with true synthetic statements, such as ($s_1$) and ($s_2$), it yields some false synthetic statements, such as "So they gave me $30 in total." Consequently, it is false that 10 + 10 = 30.

The semantic thesis is predicated on a couple of similarities between a scientific theory and a mathematical statement. Just as a scientific theory all by itself does not make any claim about observables, so a mathematical statement all by itself does not make any claims about the world. In addition, just as a scientific theory makes claims about observables once conjoined with auxiliaries, so a mathematical statement makes claims about the world once combined with some synthetic statements.

Relatedly, a mathematical sentence all by itself does not have a truth-value (Park, 2017a, p. 73). It is neither true nor false until it is supplied with synthetic statements. Once conjoined with synthetic statements, it is true or false. It is true just in case only true synthetic statements can be derived from the conjunction of it and other true synthetic statements. It is false just in case some false synthetic statements can be deduced from the conjunction of it and true synthetic statements. Therefore, it is wrong to ask whether a mathematical sentence all by itself is analytic or synthetic.

The same holds for a scientific theory. A scientific theory all by itself does not have an observational consequence, so it is wrong to ask whether it is empirically adequate until it is supplied with auxiliaries. Once conjoined with auxiliaries, it is either empirically adequate or inadequate. It is empirically adequate just in case only true observational consequences can be deduced from the conjunction of it and its true auxiliaries. It is empirically inadequate just in case some false observational consequences can be derived from it and its true auxiliaries.

There is a respect in which scientific theories and mathematical statements are dissimilar to each other. The truth of a scientific theory might or might not coincide with the empirical adequacy of it, but the truth of a mathematical statement coincides with the synthetic adequacy of it. To put it differently, some empirically adequate scientific theories are true, while other empirically adequate theories are false. By contrast, all synthetically adequate mathematical sentences are true, and none of synthetically adequate mathematical sentences is false.

The fourth is the metaphysical thesis that there is no such thing as the abstract world. It is not the abstract world but the concrete world that renders a mathematical statement true, i.e., synthetically adequate. After all, a mathematical sentence is true when its synthetic consequences are all true. The synthetic consequences are rendered true not by the abstract world but by the concrete world. Therefore, it is, in the end, the concrete world that renders a mathematical statement true.

The fifth is the epistemological thesis that we come to know that a mathematical statement is true by confirming its synthetic consequences. Using perception, for example, we can confirm ($s_1$), ($s_2$), and ($s_3$). As a result, we acquire the mathematical knowledge that 10 + 10 = 20. We do not need to confirm all of the synthetic consequences to infer that the mathematical sentence is true. In this respect, a mathematical sentence is similar to a scientific theory. We do not need to confirm all observational consequences of T to infer that it is empirically adequate; we simply ascertain some observational consequences and then infer that T is empirically adequate. Similarly, we confirm some synthetic consequences of a mathematical statement and then infer that it is synthetically adequate.

This inferentialist account of how we attain mathematical knowledge implies that what mathematical inferentialists mean by mathematical knowledge is very different from what mathematical realists mean by mathematical knowledge. For mathematical realists, mathematical knowledge is a correct mental representation of the abstract world, i.e., it is a mental state that matches up with the abstract world. For mathematical inferentialists, by contrast, to say that we know that 10 +10 = 20 means that we know that only true synthetic statements can be deduced from the conjunction of the mathematical sentence and other true synthetic statements, i.e., that we know that the mathematical sentence is a perfect inference-facilitator, an inference-facilitator that facilitates the deductive inferences from some true synthetic statements only to other true synthetic statements. In sum, to attain mathematical knowledge is to attain knowledge about the logical relationship between synthetic statements.

So far, I have clarified what mathematical inferentialism asserts. Why should we accept it? There are several reasons for choosing it over mathematical realism (Park, 2017a, pp. 74–75). Let me summarize them here. (i) Mathematical inferentialism explains why the history of science witnessed no refutation of a purely mathematical statement. It says that all purely mathematical sentences were perfect inference-facilitators, i.e., none of them yielded a false synthetic statement when conjoined with true synthetic statement. (ii) Mathematical inferentialism establishes that mathematics is an objective enterprise without postulating the existence of the queer world, i.e., the abstract world. It says that a mathematical statement is true, i.e., synthetically adequate, by virtue of the way the concrete world is, so mathematics is as objective an enterprise as empirical science. (iii) Mathematical inferentialism explains why mathematical sentences are useful in empirical science. It says that they are useful because they perform the psychological function of facilitating deductive inferences between synthetic statements. (iv) Mathematical inferentialism gives an account of how we form mathematical knowledge. It says that we acquire mathematical knowledge by confirming synthetic statements.

The refutation of this mathematical inferentialist account requires a counterexample in which a mathematical statement was discarded in the history of science, despite the revelation that it was synthetically adequate and even though it performed the perfectly useful function of facilitating deductive inferences between true synthetic statements. Another counterexample would be a mathematical statement that has been retained in the history of science despite the revelation that it was synthetically inadequate and even though it facilitates deductive inferences from some true synthetic statements to false synthetic statements.

Let me move to a different issue. In addition to mathematical realism, there is another prominent position in philosophy of mathematics called "mathematical fictionalism." According to mathematical fictionalism, "(a) our mathematical sentences and theories do purport to be about abstract mathematical objects, as platonism suggests, but (b) there are no such things as abstract objects, and so (c) our mathematical theories are not true" (Balaguer, 2018). Some readers would wonder how mathematical inferentialism differs from mathematical fictionalism and why we should choose mathematical inferentialism over mathematical fictionalism.

There are similarities and differences between mathematical inferentialism and mathematical fictionalism (Park, 2017a, p. 79). A noteworthy similarity is that they both deny that mathematical entities exist. A salient difference is that mathematical fictionalism attributes only falsity to mathematical sentences, whereas mathematical inferentialism attributes truth to some mathematical sentences and falsity to others. True mathematical sentences are those that are synthetically adequate, while false mathematical sentences are those that are synthetically inadequate. Another prominent difference is that mathematical fictionalism takes a mathematical sentence to be a description of an abstract object, whereas mathematical inferentialism does not.

Mathematical fictionalism has a disadvantage, viz., it "cannot explain why mathematics is useful in empirical science" (Park, 2017a, p. 79). This disadvantage stems from its semantic thesis that mathematical sentences describe the abstract world. It is mysterious how the descriptions of abstract objects are useful in the enterprise that describes concrete objects. On this count, mathematical fictionalism is no better off than mathematical realism and logical positivism. Only mathematical inferentialism adequately explains why mathematical sentences are useful in empirical science.

Moreover, mathematical fictionalists owe us an account of the difference between those mathematical statements that we ordinarily take to be true and those that we ordinarily take to be false. For example, what is the difference between $2 + 2 = 4$ and $1 - 1 = 2$? According to Mark Balaguer, the former is *fictionalistically correct*, while the latter is fictionalistically incorrect. A mathematical statement is "fictionalistically correct if and only if it would have been true if there had actually existed abstract mathematical objects of the kinds that platonists have in mind" (Balaguer, 2018). The abstract world, if it did exist, would make it true that $2 + 2 =4$ and would make it false that $1 - 1 =2$. It follows that "2+2=4" is fictionalistically correct, while "1-1=2" is fictionalistically incorrect.

This account of the difference between $2 + 2 = 4$ and $1 - 1 = 2$, however, runs into an epistemological objection. How do we know that the abstract world, if it did exist, would make it true that $2 + 2= 4$ and would make it false that $1 - 1 = 2$? These modal facts are causally disconnected from us, just as abstract facts are causally disconnected from us. Consequently, mathematical fictionalists owe us an account of how we know that a mathematical statement is fictionalistically correct or incorrect. It is inappropriate for them to say that we know a priori that a mathematical statement is fictionalistically correct or incorrect. After all, saying so opens the door for mathematical realists to say that we know a priori that a mathematical statement is true.

An additional objection to Balaguer's account can be raised as follows. Are Euclidean geometry, Lobachevskian geometry, and Riemannian geometry fictionalistically correct or incorrect? Mathematical fictionionalists might suggest that they are all fictionalistically correct. If Euclidean, Lobachevskian, and Riemannian worlds were to exist, they would make their respective geometries true. A problem with this account is that the Euclidean world makes Euclidean geometry true, but it makes the other two geometries false. As such, Euclidean geometry would be fictionalistically correct, but the other two geometries would be fictionalistically incorrect. However, it is not clear why we should choose the Euclidean world over the other two worlds to determine whether the other two geometries are fictionalistically correct or incorrect.

The source of this problem with mathematical fictionalism is that it invokes the abstract world to explain the difference between the mathematical statements that we ordinarily take to be true and those that we ordinarily take to be false. Of course, mathematical fictionalists are not committed to the existence of the abstract world; however, they still invoke it to explain the difference. By invoking this world, they reopen the problem that we found in Subsect. 9.3.3 of this chapter, where I argued that if it is legitimate to invoke the abstract world, it should also be legitimate to invoke a third world, a fourth world, and so forth. As a result, it becomes unclear which of these worlds we should choose to determine whether a statement is true.

## 9.8 Summary

It is a fallacious inference that mathematical objects are abstracta because attributing temporal and spatial predicates to them results in puzzling sentences. The abstract world is queer, allowing for contradictory states of affairs. The notion of a tricle undermines the realist idea that the abstract world is the truth-maker for mathematical statements. Mathematical realists have in recent decades eliminated two properties from the pool of the five properties traditionally attributed to mathematical objects; therefore, they have the burden of specifying the defining characteristics of a mathematical object.

It is to multiply cognitive faculties and entities beyond necessity for rationalists to say that the cognitive faculty of reason enables us to grasp mathematical objects and that the justification for our mathematical beliefs comes from the cognitive faculty of reason. There should be empirical evidence for causal platonism that human brains undergo changes due to mathematical objects, given that human brains are empirical objects. It is to beg the question against mathematical realists to say that we know a priori that causal platonism is true.

The indispensability argument, if correct, shows that mathematical objects exist, but it does not show that they are abstracta. It is problematic to reject the distinction between analytic and synthetic sentences yet to accept the distinction between abstract and concrete sentences. We have more evidence for the existence of interactive theoretical entities than for the existence of mathematical entities, and thus it is not intellectually dishonest to believe the existence of the former, but not the existence of the latter. Indispensabilists commit the consequential fallacy when they suggest that we negate concrete beliefs, not mathematical beliefs, because it is more convenient to do so. The consequential fallacy calls into question all beliefs in our belief system, including mathematical beliefs.

According to mathematical inferentialism, a mathematical statement promotes deductive inferences from some synthetic statements to other synthetic statements, just as a scientific theory promotes deductive inferences from auxiliaries to observational consequences. Mathematical inferentialism is more virtuous than mathematical realism and fictionalism. Therefore, scientific realists should choose mathematical inferentialism over mathematical realism and fictionalism.

## References

Baker, A. (2005). Are there genuine mathematical explanations of physical phenomena? *Mind, 114*(454), 223–237.

Baker, A. (2009). Mathematical explanation in science. *The British Journal for the Philosophy of Science, 60*(3), 611–633.

Baker, A. (2012). Science-driven mathematical explanation. *Mind, 121*(482), 243–267.

Baker, A. (2016). Parsimony and inference to the best mathematical explanation. *Synthese, 193*(2), 333–350.

Balaguer, M. (2018). Fictionalism in the philosophy of mathematics. In E. N. Zalta (Ed.), *The Stanford encyclopedia of philosophy*. https://plato.stanford.edu/archives/fall2018/entries/fictionalism-mathematics/

Benacerraf, P. (1973). Mathematical truth. *The Journal of Philosophy, 70*(19), 661–679.

Busch, J. (2012). The indispensability argument for mathematical realism and scientific realism. *Journal for General Philosophy of Science, 43*(1), 3–9.

Callard, B. (2007). The conceivability of platonism. *Philosophia Mathematica, 15*(3), 347–356.

Callard, B. (2018). Can math move matter? *Inquiry: An Interdisciplinary Journal of Philosophy.* https://doi.org/10.1080/0020174X.2018.1533886

Colyvan, M. (2001). *The indispensability of mathematics*. Oxford University Press.

Colyvan, M. (2006). Scientific realism and mathematical nominalism: A marriage made in hell. In C. Cheyne & J. Worrall (Eds.), *Rationality and reality: Conversations with Alan Musgrave* (pp. 225–237). Springer.

Duhem, P. (1905/1954). *The aim and structure of physical theories*. Trans. P. Wiener. : Princeton University Press.

Field, H. (1980). *Science without numbers*. Princeton University Press.

Hacking, I. (1983). *Representing and intervening: Introductory topics in the philosophy of science*. Cambridge University Press.

Kuhn, T. (1962/1970). *The structure of scientific revolutions*. University of Chicago Press.

Linnebo, Ø. (2018). Platonism in the philosophy of mathematics. In E. N. Zalta (Ed.), *The Stanford encyclopedia of philosophy*. https://plato.stanford.edu/archives/spr2018/entries/platonism-mathematics/.

Maddy, P. (1992). Indispensability and practice. *The Journal of Philosophy, 89*(6), 275–289.

Park, S. (2016). Against mathematical convenientism. *Axiomathes, 26*(2), 115–122.

Park, S. (2017a). In defense of mathematical inferentialism. *Analysis and Metaphysics, 16*, 70–83.

Park, S. (2017b). Two criticisms against mathematical realism. *Diametros: An Online Journal of Philosophy, 52*, 96–106.

Park, S. (2018). Rejecting mathematical realism while accepting interactive realism. *Analysis and Metaphysics, 17*, 7–21.

Park, S. (2019). Can mathematical objects be causally efficacious? *Inquiry: An Interdisciplinary Journal of Philosophy, 62*(3), 247–255.

Parsons, C. (1983). *Philosophy in mathematics: Selected essays*. Cornell University Press.

Psillos, S. (1999). *Scientific realism: How science tracks truth*. Routledge.

Putnam, H. (1971). *Philosophy of logic*. Harper and Row.

Putnam, H. (1979). *Mathematics, matter and method: Philosophical papers, volume 1* (2nd ed.). Cambridge University Press.

Quine, W. (1948). On what there is. *Review of Metaphysics, 2*(5), 21–38.

Quine, W. (1951). Two dogmas of empiricism. *The Philosophical Review, 60*(1), 20–43.

Quine, W. (1980). *From a logical point of view* (2nd ed.). Harvard University Press.

Quine, W. (1992). *Pursuit of truth*. Harvard University Press.

Resnik, M. (1997). *Mathematics as a science of patterns*. Clarendon Press.

Rosen, G. (2018). Abstract Objects. In E. N. Zalta (Ed.), *The Stanford encyclopedia of philosophy*. https://plato.stanford.edu/archives/spr2018/entries/abstract-objects/.

Trefil, J., & Hazen, R. (1998). *The sciences: An integrated approach*. Wiley.

# Chapter 10
# Scientific Realism and Scientific Practice

**Abstract** Does the realism debate matter for scientific practice? Shaw attempts to justify a positive answer to this question by providing a scientific episode in which scientists run meta-inductions and historical inductions. I point out that the meta-inductions and historical inductions are different from the pessimistic induction (PI) and the selective induction (SI), the two prominent meta-inductions and historical inductions in the realism debate. I argue that there is no evidence that the PI and the SI make any difference to scientific practice, that they should not make any difference to scientific practice, but that they have intrinsic value. I also argue that realism would promote and antirealism would forestall scientific progress if scientists adopted them as their philosophical frameworks.

**Keywords** Historical inductions · Instrumentalism · Meta-inductions · Noncommittalism · Pessimistic induction · Scientific practice · Selective induction · Surrealism

## 10.1 Introduction

Does the realism debate matter for science? Does the debate make any difference to scientific practice? Can the debate make an impact on scientific policy? Does realism have positive or negative implications for scientific practice? How about antirealism? This chapter explores opposing answers to these questions. I also delineate how realists and instrumentalists can contribute to science. This chapter should be of interest to those who wonder whether the realism debate is useful to science, whether it should be useful to science, how it can be modified to be useful to science, and whether realism and antirealism are useful philosophical frameworks for scientists. Readers will find that realism is a better philosophical framework than is antirealism for scientists.

This chapter is organized as follows. In Sect. 10.2, I critically examine the scientific episode that Jamie Shaw (2018) provides to establish that the realism debate



S. Park, *Embracing Scientific Realism*, Synthese Library 445,
https://doi.org/10.1007/978-3-030-87813-9_10

matters for scientific practice. I point out that scientists' arguments that Shaw calls "meta-inductions" and "historical inductions" are different from the pessimistic induction (PI) and the selective induction (SI), the two prominent meta-inductions and historical inductions in the realism debate. In Sect. 10.3, I argue that there is no evidence that the PI and the SI make any difference to scientific practice, that they should not make any difference to scientific practice, but that they have intrinsic value. In Sects. 10.4 and 10.5, I argue that realism and antirealism are philosophical frameworks that would respectively promote and forestall scientific progress. In Sect. 10.6, I suggest some feasible ways for realists and instrumentalists to contribute to science.

## 10.2 Scientists' Meta-inductions and Historical Inductions

In 2013, mind-brain scientists formed a research team, the Human Brain Project, "to model approximately 1% of the human brain using supercomputers" (Shaw, 2018, p. 89). Computer scientists were optimistic, while neuroscientists were pessimistic, that the research team would attain this goal. Let me summarize, one by one, their opposing arguments for their optimistic and pessimistic views.

Henry Markram, a computer scientist, developed the computational model of the brain, and his model "has steadily accumulated the ability to represent and to model neuron behavior effectively" (Shaw, 2018, p. 90). Markram predicts that his model will be more successful than it is now, given that supercomputers have been doubling their computing power just about every 18 months. He says that "we will have detailed models of mammalian brains, including that of man, in the near future" (Markram, 2006, p. 159). Consequently, the research team should work, Markram proposes, under his computational model of the brain.

Rodney Douglas, a neuroscientist, objects that there are rival theories in contemporary neuroscience and that "neuroscience is not ready to be unified under one set of paradigmatic assumptions" (Shaw, 2018, p. 91). He reasons that "the mind-brain sciences should proceed in a pluralistic fashion, with many different ontologies, since there is not enough knowledge to justify one paradigm over others" (Shaw, 2018, p. 91). The status of neuroscience "makes attempts at constructing a unified theory of the brain premature and unwarranted for the time being" (Shaw, 2018, p. 92). Shaw takes Douglas's argument to be "implicitly grounded on the history of contemporary mind-brain sciences" (Shaw, 2018, p. 92). In other words, Shaw takes Douglas as saying that since neuroscience has proceeded in a pluralistic fashion in the past, it will (or should) proceed in a pluralistic fashion in the future.

Computer scientists accept, while neuroscientists reject, that we should choose Markram's computational model of the brain. What does this disagreement have to do with scientific policy? The dispute has led neuroscientists to boycott the use of the research funds available from the Human Brain Project (Shaw, 2018, p. 90).

Thus, the differing views about whether the research team should choose Markram's computational model of the brain as the paradigm has resulted in a disagreement about how the research funds should be used.

Shaw claims that "Both computer scientists and neuroscientists use the history of their own subfields to ground their realist optimism and anti-realist pessimism, respectively" (Shaw, 2018, p. 92). As a result, Shaw refers to Markram's argument and Douglas's argument as "meta-inductions" (Shaw, 2018, p. 82, 83, 89) and as "historical inductions" (Shaw, 2018, p. 92).

Admittedly, the two scientists' arguments are meta-inductions and historical inductions. Markram's argument is a meta-argument in that it is based on an observation of computationalism, not on an observation of the world. It is also a historical induction in that the premise is about the past success of computationalism while the conclusion is about its future success. Douglas's argument is also a meta-argument in that it is based on an observation of the rival theories in neuroscience, not on an observation of the world. It is also a historical argument in that its premise is about a past event, viz., how neuroscience has proceeded in a pluralistic fashion, while the conclusion concerns a possible future event that neuroscience will (or should) proceed in a pluralistic manner. Shaw naturally refers to the argument as "a meta-induction" (Shaw, 2018, p. 82, 83, 89, 93) and "a historical induction" (Shaw, 2018, p. 92).

Markram's argument and Douglas's argument, although meta-inductions and historical inductions, should not be conflated with the PI and the SI. The PI says that "Most of the past theories of science are already suspected of being false; there is presumably every reason to anticipate that current theories of science will suffer a similar fate" (Laudan, 1977, p. 126). The PI is not about the world but rather about theories, and thus it is a meta-induction. In addition, its premise is about past theories, while its conclusion is about present theories, and thus it is a historical induction. The SI says that since the working posits of obsolete theories were preserved, the working posits of current theories will also be preserved, as we noted in Subsect. 5.6.2 of Chap. 5. The SI is not about the world but rather about the preservations of theoretical claims in theory-transitions, and thus it is a meta-induction. In addition, its premise is about the preservation of some posits of past theories that has occurred in the past, while its conclusion is about the preservation of the posits of present theories that will occur in the future, and thus it is a historical induction.

How do the PI and the SI differ from Markram's argument and Douglas's argument? The premises of the PI and the SI concern *past* theories, while the premise of Markram's argument concerns computationalism, a *current* theory. It follows that Markram's argument, although a historical argument, significantly differs from the PI and the SI. How do the PI and the SI differ from Douglas's argument? Both the PI and the SI involve scientific revolutions, while Douglas's argument does not. It follows that Douglas's argument, although a historical argument, substantially differs from the PI and the SI.

Shaw contends that "Meta-inductions are used quite frequently to justify the potential value of scientific proposals" (Shaw, 2018, p. 89). His contention is true, but it should not be confounded with the contention that the PI and the SI are used to determine the value of scientific proposals. It is one thing to say that Markram and Douglas use the meta-inductions and historical inductions to support or undermine scientific proposals; it is quite another to say that scientists use the PI and the SI to support or undermine scientific proposals. If pessimists and selective realists want to establish that the PI and the SI are operative in scientific policy, they need to provide scientific episodes in which scientists run the PI and the SI to support or undermine a scientific theory, not the kind of scientific episode that Shaw offers.

Relatedly, we should be careful when interpreting Shaw's statement "that the resolution of the realism/antirealism debate has significant implications for our criteria for theory-choice and thus for science policy" (Shaw, 2018, p. 83). This statement can be erroneously interpreted as asserting that the PI and the SI have implications for theory choice and scientific policy, given that the PI and the SI are the two salient arguments in the current realism debate.

Shaw does not specify how the resolution of the realism debate would make a difference to scientific practice. He seems to think that if realists and antirealists reach agreement that an increasingly successful theory will continue to be increasingly successful, then Markram's argument will become more robust, and as a result, computationalism will be established as a paradigm for the Human Brain Project. By contrast, if realists and antirealists reach agreement that we cannot adjudicate among the competing theories, Douglas's argument will become more robust, and as a result, neuroscientists will continue to work under the incompatible theories.

However, there is no disagreement in the first place between realists and pessimists over whether an increasingly successful theory will increasingly be successful. Neither realists nor pessimists are Humean skeptics about induction. Accordingly, they would happily agree that since Markram's computational model of the brain has been increasingly successful, it will continue to be increasingly successful. Since realists and pessimists would agree that Markram's argument is inductively correct, it is problematic to say that the resolution of the realism debate would make a difference to scientific practice.

In addition, there is no disagreement in the first place between realists and pessimists over whether immature theories are believable. To use Thomas Kuhn's (1962/1970) word, different schools of thought compete with one another over fundamentals in the stage of scientific development called "prescience." Hilary Putnam's (1975) no-miracles argument (NMA) concerns theories in a later stage of scientific development called "mature science." He says that "the theories accepted in a mature science are typically approximately true" (Putnam, 1975, p. 73). The NMA is not concerned with theories in an immature science, i.e., in prescience. Contemporary neuroscience remains in prescience since there is no clear winner among the competing theories. Consequently, realists would agree with pessimists that none of the competing theories in current neuroscience is believable.

## 10.3 The PI and the SI

Given that the PI and the SI are the two salient arguments in the current realism debate, the most convincing way to show that the realism debate matters for scientific practice is to show that the PI and the SI have an impact on it. This section aims to show that there is no evidence that the PI and the SI do not make any difference to it and that they should not make any difference to it.

It is clear that science matters for the realism debate. After all, science has shaped and molded the PI and the SI. The PI is based upon an observation of how scientists behaved in the past. For example, pessimists observe that past scientists chose their theories over previous ones and then predict that future scientists would choose future theories over present ones. The SI is also based upon an observation of how scientists behaved in the past. Selective realists observe that past scientists preserved some elements of their theories and then predict that future scientists will preserve some elements of their theories.

However, it is not clear that the PI and the SI have an impact on scientific practice. Suppose that scientists refuse to entertain an alternative to a current theory as a result of embracing the PI, which implies that the alternative will also be discarded just like the current theory. Or suppose that scientists make a conscious effort to preserve success-fueling elements of a current theory when they formulate an alternative to it, as selective realism recommends. If such events occur in science, we can conclude that the PI and the SI matter in scientific practice. The onus is on pessimists and selective realists to show that such an event has occurred.

Pessimists and selective realists offer historical episodes to establish the PI and the SI. For example, Kyle Stanford (2006, p. 20) observes that the miasma theory was replaced by the germ theory, and Stathis Psillos (1999, p. 120) observes that an experimental law was preserved in the transition from the caloric to kinetic theories. Such episodes do not show that the PI and the SI had an impact on scientific practice. It is false that scientists embraced the PI and that as a result, they chose the germ theory over the miasma theory. It is also false that scientists embraced the SI and that as a result, they made a conscious effort to preserve the experimental law of the caloric theory in the kinetic theory. These historical episodes only show that scientific practices have impacted the realism debate, not the other way around.

It is not even desirable for the PI and the SI to influence the work of scientists. There are some historical episodes in which successive theories did not share theoretical assumptions. For example, the miasma theory did not retain any theoretical assumption from the humoral theory, as we saw in Subsect. 5.6.2 of Chap. 5. Let me offer another example. The Presocratics, the initiators of science, proposed various theories about the fundamental building blocks of the universe. To put it briefly, Thales claimed that water was the basic stuff of the world; Anaximander, the boundless; Anaximenes, air; Empedocles, earth, air, fire, and water; Anaxagoras, infinitely many things; Heraclitus, fire; Parmenides, the one; Pythagoras, numbers; and Democritus, atoms (Jones, 1970, pp. 1–107). The Presocratics engaged in rigorous argumentations over the ultimate reality of the universe for hundreds of years. This book need not concern itself with the details of their argumentations. An important point, though, is that there was no common property that the Presocratics attributed to the basic stuff of the world. Therefore, the Presocratics' successive theories about the basic stuff of the universe did not have any common theoretical assumptions.

Imagine that Democritus examined all the theories about the basic stuff of the world from Thales's theory to Pythagoras's theory and that he concluded that they did not share any theoretical claims. What would have happened if he had been a pessimist? He would have predicted that a new theory to be developed would ultimately be discarded in toto, just as its forerunners were discarded in toto. As a result, he would not have bothered to develop atomism. In general, "pessimists believe that the trial and error process is repetitive and futile" (Park, 2016a, p. 121). Therefore, pessimism is a useless philosophical framework for scientists, and it should have no hold over them.[1]

What would have happened if Democritus had been a selective realist? He would have looked in vain for a common theoretical constituent in his predecessors' theories. As a result, he would have despaired, thinking that a new theory to be developed would not inherit any theoretical constituents from its precursors; more importantly, he would have predicted that it would be abandoned in toto, just as its forerunners were. As a result, he would not have bothered to entertain atomism. This thought experiment illustrates that selective realism is a useless philosophical framework for scientists and that it should have no hold over them.

This historical episode regarding atomism is important. Richard Feynman claims that atomism is the most important idea in the history of science, saying that in atomism, "there is an *enormous* amount of information about the world, if just a little imagination and thinking are applied" (Feynman, 1963, p. 4). The historical episode of atomism suggests that if Democritus had been a pessimist or a selective realist, then perhaps the most crucial idea in the history of science might not have been born, and current science might look quite different.

Pessimists and selective realists might reply that the historical episode of atomism is too far removed from current science to be relevant to the realism debate over present theories. On their account, true science began in the nineteenth century. The Presocratics merely observed the world, whereas the modern scientists of the nineteenth century performed experiments and used scientific instruments to make accurate observations and measurements. Consequently, the case study of the Presocratics is irrelevant, while the pessimists' and selective realists' case studies of nineteenth-century physics are relevant, to the realism debate concerning present theories.

This reply can be countered by the fact that the scientific methods of investigation have been improving since the early nineteenth century. For example, scientists in the nineteenth century did not have electron microscopes, particle accelerators, or the Hubble Space Telescope. Therefore, true science began, one could say, in the

[1] See Park (2017, p. 216, 2018a, p. 330) for more scientific episodes that suggest that it is undesirable for scientists to become pessimists.

early twentieth century, and all of the pessimists' and selective realists' case studies from nineteenth-century physics are irrelevant to the realism debate concerning present theories.

I have argued so far that there is no evidence that the PI and the SI have an impact on scientific practice and that they should not have any impact on scientific practice. To say so, however, does not mean that the PI and the SI are worthless. The value of a philosophical debate does not lie in its contribution to a scientific debate any more than the value of a scientific debate lies in its contribution to a philosophical debate. Stated differently, just as the value of science is independent of whether it contributes to philosophy, so the value of philosophy is independent of whether it contributes to science.

We pursue many things in our daily lives, e.g., novels, poems, sports, music, and art. We do so whether or not they contribute to science. Just as they are valuable even if they do not contribute to science, so philosophy is valuable even if it does not contribute to science. It is wrong to think that every human enterprise should be evaluated in terms of how much it contributes to science.

Consider also that philosophers are engaged in various debates in other fields of philosophy, whether or not their debates contribute to their target enterprises. For example, philosophers of religion have long been engaged in debates over the problem of evil and Pascal's wager. Suppose that the debates have had no impact on believers and disbelievers, i.e., despite being exposed to the debates, believers continue to believe in God, and disbelievers continue to disbelieve in God. Even so, we would not say that the debates are worthless.

What is the intrinsic value of the debate over the PI and the SI? My answer to this question is similar to the one that philosophers of religion would give in response to the question: what is the intrinsic value of the debates over the problem of evil and Pascal's wager? They would answer that the debates can satisfy those who wonder whether the existence of evil refutes the belief that God exists or whether we should believe in God despite the lack of evidence for belief in God. The intrinsic value of the debates cannot be appreciated by those who expect that the debates would have the effect of persuading theists into atheism or atheists into theism. Similarly, the intrinsic value of the debate over the PI and the SI is that it can satisfy those who wonder whether we can infer the collapse of current theories from that of their predecessors or whether we can infer the preservation of some posits of current theories from that of some posits of past theories. The intrinsic value of the debate cannot be appreciated by those who expect that the debate would make a difference to scientific practice.

Some philosophers are skeptical about the value of the realism debate. For example, Shaw (2018, p. 83) cites Hasok Chang, who wonders "what all the fuss is about" (Chang, 2001, p. 5). Shaw's response to such philosophers is, as we have seen in Sect. 10.2 of this chapter, to argue that the realism debate matters for scientific practice. By contrast, my response to them is to say that the realism debate has intrinsic value, just as other debates in other fields of philosophy have intrinsic value.

## 10.4 Realism and Scientific Practice

Stanford argues that the debate between realists and pessimists can make "a difference to how we should go about pursuing scientific inquiry itself" (Stanford, 2015, p. 878). More specifically, realism and pessimism can have a negative impact and a positive impact, respectively, on scientific policy. If funding agencies were full of adherents to realism, they would think that existing theories are true and that alternatives to existing theories should not be accepted. As a result, they deny funding to research proposals that challenge existing theories. Stanford writes, "The realist should be perfectly happy to rule out consideration of lines of research or theoretical proposals that are radically or fundamentally at odds with existing theories, as she thinks it quite unlikely that any of these will ultimately come to be accepted in the future" (Stanford, 2015, p. 874). On his account, scientists and scientific policymakers would be conservative if they operated under realism, and their conservative attitude would inhibit the development of new theories.

Thomas Nickles (2017) makes a similar objection to realism. Realism asserts that mature fields have reached the end of their dynamic histories, such that there will only be more applications and mop-up jobs. In other words, realism asserts that "the research frontier is closed (or nearly so) for those theories and research programs that it counts as mature" and "mature fields are sterile" (Nickles, 2016, pp. 373–374). However, it is absurd to say that no scientific progress will be made in mature fields of science. Therefore, realism is absurd.

How do I respond to Stanford and Nickles? I invite them to imagine what would happen if scientists and funding agencies embraced the counter-induction for realism enunciated in Sect. 4.4 of Chap. 4. Scientists and funding agencies would happily support the development of alternatives to existing theories. After all, they would not believe that existing theories are definitely true, but that they are likely to be true. Thus, their position is compatible with the replacement of existing theories by alternatives. Moreover, the downfall of existing theories would constitute second-order evidence for alternatives, and thus alternatives would be more likely to be true than existing theories. Therefore, scientists and funding agencies would not resist but would support the advent of new theories.

Finnur Dellsén (2019) also argues that realists welcome alternatives to existing theories. Alternatives will be more likely to be true than existing theories because the former are tested against the latter. He writes, "By the realist's lights, the newly proposed alternatives (or relevant parts thereof) will be significantly more likely to be true than the previously accepted theories (or relevant parts thereof)—after all, these theories will be tested against each other by scientific observation and experimentation (as, e.g., relativity theory was tested against classical mechanics)" (Dellsén, 2019, p. 35). Thus, Dellsén and I agree that alternatives will be more plausible than existing theories. However, he and I rely on different premises. His premise is that alternatives are tested against existing theories. By contrast, my premise (Park, 2018a, p. 335) is that the downfall of existing theories constitutes second-order evidence for alternatives.

There is another reason for thinking that the counter-induction for realism has a favorable implication for scientific policy. Scientists have failures before they make achievement, as we observed in Sect. 4.5 of Chap. 4. Suppose that a research team made multiple attempts to achieve an aim in vain. If funding agencies operated under the counter-induction, they would believe that the research team would succeed in the next attempt, and thus they would keep funding the research team. By contrast, if they operated under the PI, they would believe that the research team would fail indefinitely, and thus they would stop funding this research (Park, 2017, p. 216). Counter-inductionists shudder to imagine what would have happened to all of the outstanding scientific achievements in the history of science if scientists and scientific policymakers had espoused the PI.

The argument from undiscovered evidence, which we discussed in Sect. 4.3 of Chap. 4, also has a favorable implication for scientific practice. The argument says that since T has been supported by previously undiscovered evidence, it will continue to be supported by hitherto undiscovered evidence. The argument implies that even if an existing theory has been established, there remains undiscovered evidence for it. If scientists endorsed this argument, they would be motivated to perform new experiments on an existing theory, thinking that they can still contribute to scientific progress. Therefore, the argument from undiscovered evidence would have a positive effect on scientific progress.

In 1971, physicists flew around the earth twice on fast-flying jets to ascertain the time dilation predicted by special relativity (Hafele & Keating, 1972). Imagine that the physicists were about to board the fast-flying jets to perform the experiment. If they had endorsed the argument from undiscovered evidence, they would have thought that they might discover previously undiscovered evidence reinforcing special relativity, which was already highly confirmed. By contrast, if they had embraced pessimism, they would have thought that special relativity would be discarded, just as were its predecessors. It is clear that realism and pessimism would have had a positive impact and a negative impact, respectively, on the physicists.

The argument from neighboring theories, which we discussed in Sect. 4.4 of Chap. 4, also has a positive implication for scientific practice. The argument says that since T has been supported by previously unconceived neighboring theories, it will continue to be supported by hitherto unconceived neighboring theories. Consequently, the argument encourages scientists to pioneer new neighboring fields of science and to develop theories in these new neighboring fields of science.

## 10.5 Antirealism and Scientific Practice

### *10.5.1 Pessimism*

What would happen if funding agencies operated under the PI? They would deny funding to research proposals, whether researchers aim to build upon existing theories or to replace them with alternatives. The PI asserts that scientific theories are fated to be revealed to be false and empirically inadequate. Therefore, to build upon existing theories is to build upon false and empirically inadequate ones, and to produce alternatives to them is to produce another false and empirically inadequate theories. Either way, research proposals concern false and empirically inadequate theories; as a result, funding agencies would not be motivated to support them.

In addition, if scientists embraced the PI, they would not bother to develop new hypotheses, and they would become lethargic cognitive agents. Imagine that you are a scientist and that you develop a new scientific hypothesis in a new field of science. Unfortunately, it turns out to be false and empirically inadequate. What should you do next? You might reflect upon the following three courses of action, concluding that none of them sounds reasonable to you.

The first course of action that you might contemplate is to stay in the field and develop a new hypothesis. This course of action would sound unreasonable to you if you embraced the PI, because you would think, as a pessimist, that the demise of your first hypothesis would make your second hypothesis likely to be false and empirically inadequate. As far as pessimists are concerned, "what comes after trials and errors are only other trials and errors" (Park, 2017, p. 217).

The second course of action that you might ponder is to move into another area of science in which other scientists have already tested and discarded some hypotheses. This course of action would also sound unreasonable to you if you embraced the PI, because you would think, as a pessimist, that the demise of the other scientists' hypotheses would make your prospective hypothesis likely to be false and empirically inadequate. In sum, you would think that this old field is haunted by a pessimistic specter.

The third course of action that you might deliberate is to leave your old field and pioneer another field of science. This course of action might sound reasonable to you if you embraced the PI, because you would think, as a pessimist, that your prospective hypothesis will not have a false predecessor in the untrodden field, so it will have a chance to be true. Consequently, if scientists think that the PI is correct, they would constantly pioneer new fields of research without making multiple attempts to achieve anything in old fields of research.

On close examination, even the third course of action would sound unreasonable to you if you have embraced the PI over all first ideas that we discussed in Sect. 3.4 of Chap. 3. The PI over all first ideas implies that since all of the first ideas in all human intellectual enterprises thus far have been overturned, your new hypothesis to be developed in the new field of science will be overturned. Therefore, the only option left to you as a scientist is to stay home and watch television.

Let me bring out another implication of the PI over all first ideas in relation to the argument from neighboring theories. Imagine that you are a scientist, that you have your own theory in an old field of science, and that it is currently accepted. Some of your colleagues declare that they aim to pioneer a new research field which is close to your old field. If you embraced the argument from neighboring theories, you would support their research project. After all, the new field might produce a neighboring theory that will dovetail with your theory. By contrast, if you embraced the PI over all first ideas, you would have no reason to support their research project.

You would rather say to your colleagues that their theory to be developed in the neighboring field will be overthrown. In return, they would say, "The same thing will happen to your theory!" Pessimism is not helpful in creating an amicable atmosphere among scientists.

Nickles (2017, p. 159) advocates the PI, and thus he believes that current theories and their successors will eventually be discarded. Despite supporting the PI, however, he encourages them to develop alternatives to current theories. He says that "the position I defend encourages talented investigators to be bolder, to treat the future as still open to significant changes" (Nickles, 2017, p. 161). However, scientists would wonder why pessimists encourage scientists to develop alternatives to current theories. What is pessimists' purpose in doing so? Of course, they will see their philosophical position strengthened if scientists do as they encourage them to do. However, what do scientists get out of the deal? Pessimists cannot answer that scientists will get better theories, for that answer conflicts with their prediction that the better theories will be discarded and fuels the counter-induction for the better theories.

Let me bring this point home to philosophers with the use of the PI over antirealist accounts of the success of science that was unpacked in Sect. 3.2 of Chap. 3. Antirealists have proposed nine antirealist explanations in response to Putnam's (1975) realist explanation. Seungbae Park (2014a) exposes intrinsic problems with eight of them, but not with the latest one (Lyons, 2003), and then constructs the PI over antirealist explanations, according to which it will transpire that Timothy Lyons's proposal has intrinsic problems because it transpired that its eight predecessors have intrinsic problems. What if I were to encourage antirealists to develop an alternative to Lyons's proposal despite believing that the alternative will eventually be overturned? The following three courses of action are available to antirealists, and all of them are problematic.

First, antirealists could refuse to develop an alternative to Lyons's proposal, thinking that it will fall prey to the PI over antirealist explanations. The two advantages of this course of action are that antirealists can stick to pessimism and that they can reject my prediction that Lyons's proposal will be replaced by an alternative. A disadvantage of this course of action is that they would be stuck in a situation where they do not have an acceptable alternative to the realist explanation.

Second, antirealists could develop an alternative to Lyons's proposal, hoping that the alternative will be better than its predecessors, so it will not fall prey to the PI over antirealist explanations. An advantage of this course of action is that they might develop a viable alternative to Lyons's proposal. A disadvantage of this course of action is that they would have to believe that the PI over antirealist explanations is fallacious, and if they believe that it is fallacious, they would also have to believe that the PI over scientific theories is fallacious. As a result, they would have to give up pessimism.

Third, antirealists could denounce me for encouraging them to do something that I take to be futile, i.e., to develop an alternative that I think is fated to be renounced. An advantage of this course of action is that they can reject my encouragement to develop an alternative to Lyons's proposal. A disadvantage of this course of action is that they would also have to denounce pessimists for encouraging scientists to develop alternatives to current theories. It is not clear which of the three courses of action pessimists would choose.

Similar criticisms cannot be raised against realists. Imagine that realists encourage scientists to develop alternatives to current theories on the grounds that alternatives will be better than current theories. Scientists would not blame realists for encouraging them to do something that realists take to be futile, given that realists do not take it to be futile to arrive at better theories and given that scientists know that it is not futile to arrive at better theories. From scientists' perspective, it is reasonable for realists to encourage them to develop alternatives to current theories.

### *10.5.2 Instrumentalism*

How does Stanford argue that the PI can have a good impact on scientific policy? An answer can be found in his statement that we are "in the midst of an ongoing historical process in which successful scientific accounts of various parts of nature are repeatedly replaced with even more impressive and powerful successors making fundamentally distinct claims about the constitution and/or operation of those parts of nature" (Stanford, 2015, p. 874). He also writes "that one of the most important ambitions of the scientific enterprise should be identifying and developing the fundamentally distinct and even more powerful successors that will ultimately replace even the most impressive theories of the present day" (Stanford, 2015, pp. 874–875). These two sentences indicate that according to Stanford, scientific revolutions produce *more impressive* and *powerful* theories.

What does "more impressive and powerful" mean? It might mean that alternatives reflect unobservables better than existing theories, or it might mean that alternatives are more useful instruments than existing theories (Dellsén, 2019, p. 35). In fact, Stanford advocates instrumentalism, according to which "we might use our theories for prediction, intervention, and other pragmatic purposes without believing the theoretical descriptions they offer of the natural world" (Stanford, 2006, p. 197). On close examination, however, pessimists can say neither that alternatives represent unobservables better than existing theories nor that alternatives are more useful instruments than existing theories. After all, to say so invalidates the pessimistic inference from old to new theories and helps the counter-induction from old to new theories.

Pessimists would be tempted to retreat to instrumentalism in light of my criticisms against them in Subsect. 10.5.1 of this chapter that pessimism would make scientists lethargic cognitive agents. Regarding this possible retreat, I object that pessimism and instrumentalism are independent theses. Even if instrumentalism has a beneficial impact on scientific practice, it does not follow that pessimism also does. If instrumentalists wish to believe that pessimism also has a good impact on scientific practice, they would have to refute my criticisms against pessimism in Subsect. 10.5.1 of this chapter, and they would have to show that pessimism, but not instrumentalism, has a good impact on scientific practice. To put it differently, it is one thing to say that instrumentalism has a beneficial impact on science, and it is entirely another to say that pessimism has a beneficial impact on science.

It is not even clear whether instrumentalism has a beneficial impact on scientific practice. Instrumentalism has several pragmatic and epistemic disadvantages, as we noted in Chap. 3. For example, instrumentalists cannot use scientific theories to predict and explain phenomena due to three problems: Moore's paradox, the problem of disconcerting questions, and the problem of deceptive speech acts. Therefore, instrumentalism is not a worthwhile philosophical framework for scientists. However, I set aside these problems in this chapter. Instead, I explore whether instrumentalists lend more support to scientific revolutions than do realists.

Resolving the issue of who would lend more support to scientific revolutions requires that we consider what realists and instrumentalists gain and lose from scientific revolutions and that we compare their gains and losses. If realists gain more and lose less than instrumentalists, realists would support scientific revolutions more than would instrumentalists. If instrumentalists gain more and lose less than realists, instrumentalists would support scientific revolutions more than would realists.

Realists gain and lose different beliefs, depending on what exactly they believe regarding existing theories and their alternatives; the same is true of instrumentalists (Park, 2019a, pp. 149–153). Consequently, we should consider precisely what realists and instrumentalists believe regarding existing theories and their alternatives before comparing their gains and losses. Realists might believe that existing theories and their alternatives are true, or they might believe that they are approximately so. Instrumentalists might believe that existing theories and their alternatives are useful, or they might believe that they are empirically adequate (EA). Alternatively, they might believe that they are approximately EA.

Suppose that realists and instrumentalists believe that existing theories are true and EA, respectively, not that they are approximately true and approximately EA. Who would be more supportive of the development of alternatives to existing theories? It is not clear what the correct answer to this question is. When existing theories are superseded by alternatives, realists lose their previous beliefs that existing theories are true, and they form the new beliefs that alternatives are true. As a result, they get closer to truth. By contrast, instrumentalists lose the previous beliefs that existing theories are EA, and they now believe that alternatives are EA. As a result, they get closer to empirical adequacy. In short, "Realists lose a lot but gain a lot, whereas instrumentalists lose a little but gain a little from scientific revolutions" (Park, 2019a, p. 152). Under these circumstances, it is not clear who would lend more support to scientific revolutions.

Suppose now that realists and instrumentalists believe that existing theories are approximately true and useful, respectively. After their alternatives have replaced existing theories, realists believe that both existing theories and their alternatives are approximately true and useful, whereas instrumentalists believe only that both existing theories and their alternatives are useful. Under these circumstances, both realists and instrumentalists lose nothing from scientific revolutions, i.e., realists retain their previous beliefs that existing theories were approximately true, and instrumentalists also retain their earlier beliefs that existing theories were useful. Realists, however, get closer to truth, while instrumentalists get closer to empirical adequacy. In other words, realists gain a lot, while instrumentalists gain a little, from scientific revolutions. Consequently, realists would be more supportive of scientific revolutions than would instrumentalists (Park, 2019a, p. 152).

Stanford (2015) would object that realists stretch the notion of approximate truth, arguing that alternatives are fundamentally distinct from existing theories. Realists cannot, he would continue, have it both ways. If alternatives to existing theories are approximately true, then existing theories cannot be approximately true. My response to this criticism is to point out that the criticism cuts both ways. Instrumentalists also stretch the notion of usefulness. Existing theories will be replaced by their alternatives because the latter handle the anomalies to the former, which implies that the former are empirically inadequate (Park, 2014b, p. 112; Dellsén, 2019, p. 37). Thus, existing theories will be, to some degree, useless. If it were questionable to attribute "approximately true" to existing theories, it would also be questionable to attribute "useful" to them.

Suppose now that realists believe that existing theories are approximately true and EA and that instrumentalists believe that they are approximately EA. After existing theories are replaced by their alternatives, realists believe that alternatives are approximately true and EA, and instrumentalists believe that they are approximately EA. Under these circumstances, realists lose nothing and gain a lot, while instrumentalists lose nothing and gain a little, from scientific revolutions. Consequently, realists would be more supportive of scientific revolutions than instrumentalists (Park, 2019a, p. 152).

Stanford (2006) believes that existing theories are useful, but he does not state one way or the other whether he believes that existing theories are EA or approximately EA. Instrumentalists cannot believe that existing theories are less than approximately EA. After all, if a theory is less than approximately EA, "it is not clear how it can be useful for making predictions and manipulations" (Park, 2019a, p. 152). It follows that instrumentalists would have to believe either that existing theories are EA or that they are approximately EA. Whichever they may believe, realists would be as supportive of scientific revolutions as they or more supportive of scientific revolutions than they. Therefore, it is wrong for Stanford to say that realists would be more conservative than instrumentalists.

### *10.5.3 Surrealism*

Let me turn to another popular version of antirealism called "surrealism." According to surrealism, observables operate as if T were true. On this account, T is successful because observables behave as if it were true. Note that surrealism invokes the behavioral pattern of observables to explain the success of T. As noted in Sect. 1.10 of Chap. 1, many venerable philosophers take surrealism to be a serious competitor to realism, including Arthur Fine (1986, p. 160), Jarrett Leplin (1987), Alan Musgrave (1988, p. 243), André Kukla (1998, p. 22), Stanford (2000, p. 268), Moti Mizrahi (2012, p. 133), and Lyons (2003, p. 897, 2018, p. 147).

Scientists should be creative and work hard to produce a successful theory, as we noted in Sect. 1.10 of Chap. 1. Imagine that you are a scientist, that you have devised T, and that it shows an impressive performance, i.e., it makes true predictions. You have two fellow scientists. One has embraced realism and the other surrealism. They say to you, respectively, "Your theory is successful because it is true," and "Your theory is successful because observables behave as if it were true." The realist and the surrealist provide different explanantia to explain the same explanandum, the impressive performance of T. The realist's explanans is a property of T, whereas the surrealist' explanans is the way the world behaves. On the realist's account, the explanation of the impressive performance of T lies in T, whereas on the surrealist's account, the explanation of the impressive performance of T lies in the world. This difference implies that on the realist's account, T shows the impressive performance due to your originality and hard work, whereas on the surrealist's account, T shows the impressive performance due to the world. In sum, the realist's explanation recognizes your originality and hard work, whereas the surrealist's explanation does not.

In this context, it is useful to note what a historian of science says about what drives scientific progress. Jeffrey Seeman says that "recognition of originality is a crucial aspect and driving force in the progress of science" (Seeman, 2018, p. 219). It follows that if realism and surrealism permeated scientists' minds, they would have positive and negative impacts, respectively, on scientific practice. Realists' explanation of why T is successful could create an atmosphere where scientists recognize their colleagues' achievements. By contrast, surrealists' explanation of why T is successful could create an atmosphere where scientists belittle their colleagues' achievements. Like pessimism, surrealism is not helpful in creating a friendly atmosphere among researchers.

Moreover, suppose that scientists devise a new theory, but that it fails empirical tests. Why is it unsuccessful? Scientists would give different explanations to themselves, depending on whether they subscribe to realism or surrealism. If they subscribed to realism, they would think to themselves that it is unsuccessful because it is false. This explanation blames themselves, thereby yielding a useful prescription on what they should do next, viz., they should revise it or devise a new theory. If they subscribed to surrealism, by contrast, they would think to themselves that it is unsuccessful because observables do not operate as if it were true. This explanation blames the world, thereby yielding a useless prescription on what they should do next, viz., they should change the way observables behave. In sum, the realist explanation provides a useful prescription, whereas the surrealist explanation provides a useless prescription, for scientists regarding how to come up with a successful theory in their future endeavors (Park, 2016b, pp. 610–611).

What can we conclude from these pragmatic disadvantages of surrealism? Surrealism is a useless philosophical framework for scientific practice. Surrealism is merely a philosophical toy concocted for the sole purpose of undermining realism. We might grant that antirealists succeeded in undermining realism by proposing the alternative to realism. However, their victory is a Pyrrhic one.

### 10.5.4 Noncommittalism

Antirealists might take the position called "noncommittalism." It is "the position that is not committed to any claim" (Park, 2018b, p. 10). It does not assert that T is EA; consequently, it is immune to the PI, which implies that both past and present theories are empirically inadequate. Nor is it committed to the position that T is useful; accordingly, it is immune to my critiques of instrumentalism in Subsect. 10.5.2 of this chapter. Noncommittalism seems to be a perfect philosophical position for those who are loathe to be criticized by others.

Noncommittalism is different from skepticism. Skepticism claims that we do not have a sufficient amount of evidence for a specific target of an investigation, whether it is God, a theoretical entity, the mind, a physical object, a moral property, or a mathematical object. Noncommittalism does not even claim, one way or the other, whether there is a sufficient amount of evidence for a specific target of an investigation. As a result, it is even immune to Douglas Long's (1992) criticism against Cartesian skepticism. According to Long, the mere fact that conscious mental properties exist does not show that the mind, which putatively has mental properties, exists, yet Cartesian skepticism presupposes that the mind exists, and therefore Cartesian skepticism is incoherent. Noncommittalism is silent about whether conscious mental properties exist or whether the mind exists. Therefore, Long's critique of Cartesian skepticism does not apply to noncommittalism.

However, noncommittalism has epistemic and pragmatic disadvantages (Park, 2018b, pp. 10–11). If you espouse it, you must even be noncommittal about your own positive theories, and you should not hold a grudge against your epistemic colleagues who are noncommittal about your positive theories. As a result, no one might believe your theories. Imagine that you think that your theories are so virtuous that you believe that you deserve a scholarly award for your ingenuity. You submit your application to the award committee. However, the award committee members refuse to grant you scholarly recognition because they are noncommittal about your positive theories.

It is not difficult to imagine what would happen to scientific practice if all scientists embraced noncommittalism. They would be listless and lethargic cognitive agents, not even bothering to conceive hypotheses, nor would they bother to perform experiments. After all, they would be noncommittal about their own hypotheses and experiments, and their fellow scientists would also be noncommittal about their hypotheses and experiments. Where there is no incentive, there is no science. In sum, noncommittalism would have a lethal effect on scientific practice, and it should not be a philosophical framework for scientists.

Let me draw a philosophical moral from the epistemic and pragmatic disadvantages with noncommittalism. Philosophers tend to retreat to noncommittalism or skepticism when their position is under fire. For example, I object that it is incoherent for Bas van Fraassen to reject inference to the best explanation and then use it to establish the contextual theory (Park, 2019b, p. 91). Playing devil's advocate, Richard Healey (2019, pp. 26–27) replies that van Fraassen does not believe the contextual theory as we noted in Sect. 3.7 of Chap. 3. Psillos (1997, p. 370) objects that the inference from success to empirical adequacy is ampliative. Brad Wray (2012, p. 378) replies that constructive empiricists do not believe that T is EA. Admittedly, to fall back to noncommittalism or skepticism is an effective and standard move to fend off criticisms. However, this effective and standard move incurs epistemic and pragmatic disadvantages, and hence it brings at best a Pyrrhic victory.

## 10.6 Contribution

In the immediately previous section, I discussed four variants of antirealism: pessimism, instrumentalism, surrealism, and noncommittalism. Of these four, instrumentalism is the most promising. In this section, I suggest some possible ways for instrumentalists and realists to contribute to science.

Realists can propose new scientific theories to scientists. At first blush, it appears incoherent for realists to offer new theories. On close examination, it is not. As noted in Sect. 10.4 of this chapter, realists welcome new theories to displace old ones because they believe that new ones will be more likely to be true than old ones. Moreover, new theories might explain and predict phenomena in new fields of research, or they might serve as neighboring theories reinforcing existing theories in old fields of research.

In addition, realists can propose new experiments for scientists. The new experiments, if performed by scientists, might support or undermine existing theories. If existing theories are supported by new experiments, they would be more likely to be true. Realists welcome existing theories becoming more likely because they believe that they are true. What if existing theories are undermined by the experiments? Realists also welcome such a consequence because it might prove to be an opportunity to develop new theories and because realists believe that new theories will be more probable than existing ones.

Instrumentalists can also propose new scientific theories to scientists. Instrumentalists support the replacement of existing theories with alternatives because they believe that alternatives will be better instruments than existing theories, or the new scientific theories might prove to be useful instruments in new areas of research. Alternatively, existing theories might predict new phenomena once they are conjoined with new theories, i.e., new theories might serve as neighboring theories that expand the use of existing theories. Instrumentalists welcome such consequences of new theories.

In addition, instrumentalists can propose new experiments for scientists. The results of the new experiments might agree or disagree with existing theories. To agree with existing theories means that the use of existing theories is extended. Instrumentalists welcome existing theories becoming more useful. To disagree with existing theories implies that existing theories are useless for certain phenomena, but such a consequence might prove to be an opportunity to develop better instruments, and instrumentalists welcome the development of better instruments.

I propose that realists and instrumentalists change the targets of their case studies. They have been presenting the case studies of past theories, especially physical theories of the nineteenth century, to ascertain the fate of present theories. However, predicting the fate of present theories requires not the case studies of past theories but rather the case studies of present theories. As they dig into the scientific literature to come up with the case studies of present theories, they might hit upon new theories and experiments.

I do not propose that instrumentalists and realists build scientific instruments, such as particle accelerators and the Hubble Space Telescope, to contribute to scientific practice. To construct such scientific instruments would clearly count as contributions to science, but that would require that philosophers change their careers. Only professional scientists and engineers can make such contributions to science. To require that philosophers make such contributions to science is just as unreasonable as to expect that scientists publish papers in prestigious philosophy journals. Scientists would have to change their careers to make such contributions to philosophy.

Just as it is unreasonable to require that scientists should have as much philosophical knowledge as philosophers to make a difference to the realism debate, so it is unreasonable to require that philosophers should have as much scientific knowledge as scientists to contribute to scientific practice. Philosophers and scientists have different expertise, although they might have some overlapping interests and although they might interact with one another under the common research agenda.

There are more feasible ways for instrumentalists and realists to contribute to scientific practice: it is to teach philosophy of science courses to science majors and to publish papers and/or books accessible to scientists. Science majors will be exposed to how great scientists made their achievements. As a result, they may get an idea what they should do to become like them. Also, they will learn that pessimism, surrealism, and noncommittalism have harmful effects on science and that realism has a more beneficial impact on science than instrumentalism.

## 10.7 Summary

There is no evidence that the PI and the SI affect science, and they should not affect science. If scientists operated under the PI, they would think that after trials and errors, there will only be more trials and errors. Thus, they would think that it is futile to develop an alternative to an existing theory. If they operated under the SI, they would think that it is pointless to develop an alternative to an existing theory that did not inherit any assumptions from its predecessor. However, the SI and the PI have intrinsic value.

Realism is better than antirealism as a philosophical framework for scientific progress. If realists and instrumentalists aim to contribute to scientific practice, they should set aside the PI and the SI, present case studies of current theories, teach philosophy of science courses to science majors, and/or publish papers and books useful to scientists. Let me emphasize, though, that the value of philosophy of science does not lie in its contribution to science any more than the value of science lies in its contribution to philosophy of science.

## References


Chang, H. (2001). How to take realism beyond foot-stomping. *Philosophy, 76*(1), 5–30.

Dellsén, F. (2019). Should scientific realists embrace theoretical conservatism? *Studies in History and Philosophy of Science, 76*, 30–38.

Feynman, R. (1963). *Six easy pieces: Essentials of physics explained by its most brilliant teacher*. Basic Books.

Fine, A. (1986). Unnatural attitudes: Realist and instrumentalist attachments to science. *Mind, 95*(378), 149–179.

Hafele, J., & Keating, R. (1972). Around-the-world atomic clocks: Predicted relativistic time gains. *Science, 177*(4044), 166–168.

Healey, R. (2019). The aims of reliable knowledge: Reply to Seungbae Park. *Social Epistemology Review and Reply Collective, 8*(9), 25–30.

Jones, T. (1970). *A history of Western philosophy: The classical mind*. Wadsworth.

Kuhn, T. (1962/1970). *The structure of scientific revolutions*. University of Chicago Press.

Kukla, A. (1998). *Studies in scientific realism*. Oxford University Press.

Laudan, L. (1977). *Progress and its problems: Towards a theory of scientific growth*. University of California Press.

Leplin, J. (1987). Surrealism. *Mind, 97*(384), 519–524.

Long, D. (1992). The self-defeating character of skepticism. *Philosophy and Phenomenological Research, 52*(1), 67–84.

Lyons, T. (2003). Explaining the success of a scientific theory. *Philosophy of Science, 70*(5), 891–901.

Lyons, T. (2018). Four challenges to epistemic scientific realism. *Spontaneous Generations: A Journal for the History and Philosophy of Science, 9*(1), 146–150.

Markram, H. (2006). The blue brain project. *Nature Reviews Neuroscience, 7*(2), 153–160.

Mizrahi, M. (2012). Why the ultimate argument for scientific realism ultimately fails. *Studies in History and Philosophy of Science Part A, 43*(1), 132–138.

Musgrave, A. (1988). The ultimate argument for scientific realism. In R. Nola (Ed.), *Relativism and realism in science* (pp. 229–252). Kluwer Academic Publishers.

Nickles, T. (2016). Perspectivism versus a completed Copernican revolution. *Axiomathes, 26*(4), 367–382.

Nickles, T. (2017). Cognitive illusions and nonrealism: Objections and replies. In E. Agazzi (Ed.), *Varieties of scientific realism: Objectivity and truth in science* (pp. 151–163). Springer.

Park, S. (2014a). A pessimistic induction against scientific antirealism. *Organon F, 21*(1), 3–21.

Park, S. (2014b). Approximate truth vs. empirical adequacy. *Epistemologia, 37*(1), 106–118.

Park, S. (2016a). How to foster scientists' creativity. *Creativity Studies, 9*(2), 117–126.

Park, S. (2016b). Realism versus surrealism. *Foundations of Science, 21*(4), 603–614.

Park, S. (2017). The uniformity principle vs. the disuniformity principle. *Acta Analytica, 32*(2), 213–222.
Park, S. (2018a). The anti-induction for scientific realism. *Grazer Philosophische Studien, 95*(3), 329–342.
Park, S. (2018b). The problem of unobserved anomalies. *Filosofija. Sociologija, 29*(1), 4–12.
Park, S. (2019a). Should scientists embrace scientific realism or antirealism? *Philosophical Forum, 50*(1), 147–158.
Park, S. (2019b). The disastrous implications of the 'English' view of rationality in a social world. *Social Epistemology, 33*(1), 88–99.
Psillos, S. (1997). How not to defend constructive empiricism: A rejoinder. *The Philosophical Quarterly, 47*(188), 369–372.
Psillos, S. (1999). *Scientific realism: How science tracks truth.* Routledge.
Putnam, H. (1975). *Mathematics, matter and method: Philosophical papers, volume 1.* Cambridge University Press.
Seeman, J. (2018). From 'multiple simultaneous independent discoveries' to the theory of 'multiple simultaneous independent errors': A conduit in science. *Foundations of Chemistry, 20*(3), 219–249.
Shaw, J. (2018). Why the realism debate matters for science policy: The case of the human brain project. *Spontaneous Generations: A Journal for the History and Philosophy of Science, 9*(1), 82–98.
Stanford, K. (2000). An antirealist explanation of the success science. *Philosophy of Science, 67*(2), 266–284.
Stanford, K. (2006). *Exceeding our grasp: Science, history, and the problem of unconceived alternatives.* Oxford University Press.
Stanford, K. (2015). Catastrophism, uniformitarianism, and a scientific realism debate that makes a difference. *Philosophy of Science, 82*(5), 867–878.
Wray, B. (2012). Epistemic privilege and the success of science. *Noûs, 46*(3), 375–385.

# Chapter 11
# New Topics for Future Debates

**Abstract** Stanford makes two proposals to revamp the realism debate. The first proposal is that rival discussants stop debating whether our best theories are approximately true. I object that this proposal overlooks the fact that there are realists in the literature in addition to pessimists and selective realists. The second proposal is that rival disputants start debating whether current theories will be superseded by alternatives, modeling the realism debate on the geological debate between catastrophists and uniformitarians. I object that it is self-refuting for a pessimist to appeal to a scientific debate. Finally, I introduce several new topics for future debates between realists and antirealists, extracting most of them from previous chapters.

**Keywords** Double spaces · Counter-induction · Neighboring theories · Pessimistic induction · Selective induction · Scientific revolutions · Undiscovered evidence · Unvirtuous prediction

## 11.1 Introduction

As noted in Chap. 10, the two salient arguments in the current realism debate are the pessimistic induction (PI) and the selective induction (SI). There is a problem with the present status of the realism debate. The gap between the PI and the SI is so narrow that there cannot be much debate between their advocates; as Kyle Stanford (2015) observes, there is only a verbal dispute between them. Stanford (2015, 2018) makes the following two proposals to revamp the realism debate. First, rival discussants should stop talking about whether our best theories are approximately true. Second, rival disputants should start talking about whether there will be scientific revolutions, modeling the realism debate on the geological debate between catastrophists and uniformitarians. In this chapter, I make critical comments about these two proposals and make alternative proposals.

This chapter is organized as follows. In Sect. 11.2, I unpack Stanford's motivations for putting forth the two proposals. I criticize his first proposal, pointing out that he overlooks the fact that there are realists in the literature. Realists' attribution


S. Park, *Embracing Scientific Realism*, Synthese Library 445,
https://doi.org/10.1007/978-3-030-87813-9_11

of the phrase "approximately true" to present theories is predicated not only on the no-miracles argument (NMA) but also on the six new arguments for realism that we have discussed in Chap. 4. I also criticize the second proposal, pointing out that it is self-defeating for a pessimist to propose that the realism debate is to be modeled on the scientific debate. In Sect. 11.3, I introduce several new topics for future debates between realists and antirealists. This chapter is intended to be useful to those who seek new topics concerning realism.

## 11.2 Stanford's Two Proposals

### *11.2.1 Motivation*

As we have seen in Subsection 6.1 of Chap. 5, pessimists and selective realists share the prediction that current theories will follow the course of their forerunners, and there is only a terminological dispute between them. Selective realists attribute, while pessimists refuse to attribute, "approximately true" to past and current theories. The terminological dispute cannot be resolved by rational argumentation. Consequently, selective realists and pessimists have reached a stalemate.

Observing this stalemate, Stanford makes two insightful proposals. The first is that participants in the realism debate "simply stop thinking, talking, and writing about the scientific realism debate as being concerned with the approximate truth of our best scientific theories" (Stanford, 2018, p. 80). The second is that participants instead start debating whether present theories will be supplanted by alternatives. Under this new framework, realists would argue that current theories will be retained, while pessimists would argue that they will be discarded.

The newly structured debate between realists and pessimists parallels the debate between catastrophists and uniformitarians in geology (Stanford, 2015, pp. 875–876, 2018, pp. 78–79). In geology, uniformitarianism holds that natural events, such as earthquakes, floods, and volcanoes, have caused radical geological changes and that they will bring about geological changes "just as profoundly as they have been" (Stanford, 2015, p. 876). By contrast, catastrophism holds that such radical geological changes are "confined to the distant past" (Stanford, 2015, p. 876) and that there will only be small geological changes from now on.[1] Just as catastrophists believe that there were geological upheavals in the past but there will not be anymore, so realists believe that there were radical changes in theory but there will not be anymore. Similarly, just as uniformitarians assume that geological upheavals will continue to occur as they have before, so pessimists believe that there will be continue to be radical theory changes as there have before. An advantage of restructuring the

[1] Stanford seems to have misunderstood catastrophism and uniformitarianism. Catastrophism claims that geological changes are abrupt and radical, whereas uniformitarianism claims that geological changes are slow and gradual. However, I set aside this issue and grant that Stanford is right about catastrophism and uniformitarianism.

debate between realists and pessimists in this way is that the debate can dispense with the terminological issue of whether it is legitimate to attribute "approximately true" to present theories. Instead, the debate can focus on the substantive issue of whether there will be scientific revolutions.

Now that we are clear about what motivates Stanford to make the two proposals, we are ready to evaluate them. I evaluate the first proposal in Sect. 11.2.2 and the second proposal in Sect. 11.2.3 below.

### *11.2.2 Responses to Stanford's First Proposal*

How do I respond to Stanford's first proposal that the disputants in the realism debate should stop using the phrase "approximately true"? This proposal is based upon the assumption that there are only pessimists and selective realists in the literature. This assumption is reasonable. As we noted in Subsection 6.1 of Chap. 5, selective realism is in vogue today. Contrary to this assumption, however, there are still realists. They do not believe that scientific revolutions will occur, and thus they are not stretching the notion of approximate truth when they apply "approximately true" to present theories. In addition, they rely not only on the NMA but also on the six new arguments for realism that are unpacked in Chap. 4.

There is no doubt that pessimists would criticize the six new arguments for realism, insisting that there is no rational basis for attributing "approximately true" to present theories. However, my point is not that the six new arguments are promising, but rather that Stanford's first proposal requires them to be refuted first. Moreover, realists are not merely expressing their taste when they attribute "approximately true" to present theories. In short, the dialectical terrain is not as simple as Stanford's first proposal presupposes.

Stanford has a point, though. In his spirit, I propose that we stop attaching the name "realism" to the various positions developed under the framework that present theories will undergo scientific revolutions. "Selective realism" and "structural realism" are cases in point.[2] These names give the impression that they refer to variants of realism. As Stanford points out, however, it is controversial whether the components of present theories to which they are committed are rich and thick enough to entitle selective and structural realists to attribute "approximately true" to present theories.

[2] See Chap. 5 for other positions to which "realism" is questionably attached.

### 11.2.3 Responses to Stanford's Second Proposal

What are we to make of Stanford's second proposal that the realism debate is to be modeled on the geological debate? It is self-defeating for a pessimist to make such a proposal. Given that uniformitarianism is a scientific theory, pessimism implies that it is fated to be replaced by a radically distinct alternative. Since it was born with this sad fate, its philosophical counterpart, viz., pessimism, was also born with the sad fate, i.e., it is fated to be replaced by a radically distinct philosophical alternative.

Uniformitarianism and pessimism should be treated equally. Suppose that catastrophists and uniformitarians have heated debates and that uniformitarianism eventually comes out the winner. It does not follow that uniformitarianism is true. After all, it will in turn be ousted by a radically distinct alternative. Consequently, uniformitarianism is false. Similarly, suppose that realists and pessimists have their heated debates and that pessimism finally comes out the winner. Just like before, it does not follow that pessimism is true. After all, it will be ousted by a radically distinct alternative. Consequently, pessimism is false.

As we noted in Subsection 4.2 of Chap. 2, Stanford contends that there are indefinitely or infinitely many alternatives, saying that the space of alternatives "appears to be indeterminate and unbounded" (Stanford, 2006, p. 133). Scientists can never eliminate all of the false alternatives that need to be eliminated before they arrive at a true scientific theory. They are "creatures whose cognitive constitutions are not well suited to the task of exhausting the kinds of spaces of serious candidate theoretical explanations from which our scientific theories are drawn" (Stanford, 2006, p. 45). This is quite a remarkable argument against the optimistic view that scientists can obtain true scientific theories.

This chapter grants that Stanford is right about the number of unconceived alternatives and then runs a similar argument against Stanford's proposal that the realism debate is to be framed on the geological debate. Given that uniformitarianism is a scientific theory, it competes with infinitely many unconceived scientific alternatives. It follows that pessimism also competes with infinitely many unconceived philosophical alternatives. Just as scientists cannot exhaust the possibility space of infinitely many scientific alternatives, pessimists cannot exhaust the possibility space of infinitely many philosophical alternatives. Thus, pessimists are in a situation in which they are not sure that pessimism is true.

Stanford cannot have it both ways. On the one hand, he accepts pessimism, which implies that scientific theories, including uniformitarianism, are false. On the other hand, he claims that pessimism and uniformitarianism make similar claims, noting that pessimism asserts that there will be radical theoretical changes as there were before and that uniformitarianism claims that there will be radical geological changes as there were before. If he accepts pessimism, however, he must reject uniformitarianism because pessimism implies that uniformitarianism is false. Thus, pessimism backfires on his own positive philosophical proposal that the realism debate be modeled on the geological debate.

How would Stanford fend off this criticism? He might argue that pessimism and uniformitarianism are dissimilar, since pessimism was advanced by philosophers while uniformitarianism was offered by scientists. Philosophers have, while scientists do not have, the cognitive capacity to exhaust the possibility space of alternatives. It follows that pessimism does not compete with infinitely many unconceived philosophical alternatives, but uniformitarianism does compete with infinitely many unconceived scientific alternatives. Therefore, uniformitarianism is, while pessimism is not, fated to be replaced by alternatives.

An immediate objection to this move is that if there is such a profound difference between uniformitarianism and pessimism, it is wrong to propose that the realism debate is to be framed on the geological debate. There should be some clarification as to how one can claim both that uniformitarianism and pessimism have the aforementioned profound difference and that the realism debate should be structured like the geological debate.

Moreover, it is arbitrary to suggest that philosophers have, whereas scientists do not, the cognitive capacity to exhaust the possibility space of alternatives. This suggestion merely manifests philosophical arrogance and applies a double standard to philosophers and scientists. Sharp scientists would reverse the double standard, saying that scientists have, whereas philosophers do not, the cognitive capacity. This possible response from astute scientists should remind philosophers of the Golden Rule. If philosophers do not want scientists to assert that scientists have, but philosophers do not have, the cognitive capacity to exhaust the possibility space of alternatives, philosophers should not assert that philosophers have, but scientists do not have, the cognitive capacity. There is "no reason for thinking that the Golden Rule ranges over moral matters, but not over epistemic matters" (Park, 2018, pp. 77–78).

Pessimists might now argue that the analogy between the realism debate and the geological debate may be stretched as little or as much as needed. After all, the takeaway is not that pessimists should be uniformitarians but that realists and pessimists should focus on the substantive issue of whether or not there will be scientific revolutions. Stated differently, Stanford can propose that realists and pessimists focus on the substantive issue aside from the geological debate.

Let me make a couple of comments on this possible move. First, it amounts to giving up Stanford's second proposal that "we should reconceive the most important locus of disagreement in the debate concerning scientific realism along the lines of the great clash between catastrophism and uniformitarianism in nineteenth-century geology" (Stanford, 2015, pp. 875–876). There is a cost to jettisoning the analogy between the geological debate and the realism debate, viz., it would be wrong for pessimists to investigate the details of the geological debate, whatever those are, intending to flesh out the future direction of the realism debate.

Second, this potential move presupposes that there are only two positions available to the disputants of the realism debate. One position is that there will be scientific revolutions, and the other position is that there will not be any scientific revolutions. However, this presupposition is false, given that there are profoundly distinct philosophical alternatives to these two positions. They are unconceived, but they will be conceived. Moreover, such alternatives are infinitely many. Note that these claims are independent of the geological theories.

Pessimists might argue that they cannot ideate philosophical alternatives to pessimism and realism, so there are no such alternatives. However, this argument only invites the reply that they cannot ideate the alternatives not because there are no such alternatives but rather because their cognitive capacities are limited. This reply mirrors pessimists' contention that scientists cannot think of alternatives to current theories not because there are no such alternatives but rather because their cognitive capacities are limited.

Critics might now object that I have not provided an inductive argument for the existence of philosophical alternatives to realism and pessimism. My position contrasts with Stanford's position. Stanford (2006, pp. 19–20) has provided an inductive argument for the existence of unconceived alternatives to present scientific theories. He puts forward a list of past and current theories and then concludes that there are unconceived alternatives to present theories, just as there were unconceived alternatives to past theories. However, I have not provided any such list. In its absence, it is merely an assumption that there are infinitely many philosophical alternatives to realism and pessimism.

In response to this objection, one might make a list of philosophical theories that were accepted in the past but are rejected now. The list would include such philosophical theories as Aristotle's teleological view of the world and René Descartes's substance dualism. Based on such a list, one could infer that there are alternatives to realism and pessimism. Such a strategy mirrors that of pessimists.

In this context, however, let me try a different move: constructing the following argument from analogy. Pessimism and uniformitarianism are similar in that both assert that there will be dramatic changes in the future, just as there have been in the past. Specifically, pessimism claims that there will be scientific revolutions in the future just as there have been in the past, and uniformitarianism claims that there will be geological upheavals in the future just as there have been in the past. Uniformitarianism competes with infinitely many scientific alternatives. Therefore, pessimism also competes with infinitely many philosophical alternatives.

The moral is that the analogy between pessimism and uniformitarianism exists, even if Stanford does not make use of it. Moreover, his critics can make use of it to refute pessimism, not to support it. As long as pessimists stick to the idea that there are unconceived scientific alternatives to each scientific theory, they will face the objection that there are unconceived philosophical alternatives to pessimism. This section can be summed up in this slogan: "A cognitive famine lurks behind pessimism."

So far, I claimed that it is self-refuting for pessimists to propose that the realism debate be modeled on the geological debate. However, my claim does not entail that it is an illegitimate topic whether there will be scientific revolutions in the future. In fact, I believe that the subject is a legitimate one. In the next section, I introduce several other topics for future debates between realists and antirealists, developing most of them from the topics in the previous chapters.

## 11.3 Future Topics

### *11.3.1 Resolving Conflicts*

The PI conflicts with the argument from undiscovered evidence, with the argument from neighboring theories, and with the counter-induction for realism that we discussed in Chap. 4. How can we resolve these conflicts? I leave the task of answering this question to future researchers.

The PI asserts that since earlier theories were displaced by previously unconceived alternatives, their successors will continue to be displaced by hitherto unconceived alternatives. If the PI is correct, current theories are false. To reiterate, the PI does not merely assert that current theories are unwarranted. Stanford says, for instance, that "the scientific theories of the past have turned out to be false despite exhibiting just the same impressive sorts of virtues that present theories do, so we should expect our own successful theories to ultimately suffer the same fate" (Stanford, 2006, p. 7). To suffer the same fate is to turn out to be false.

The PI conflicts with the argument from undiscovered evidence, according to which since T has been supported by previously undiscovered evidence, it will continue to be supported by hitherto undiscovered evidence. If this argument is correct, T is true; however, the PI implies that T is false. We can resolve this conflict between the PI and the argument from undiscovered evidence by showing that either or both are incorrect. I do not rule out the possibility that both are incorrect. It will be interesting to see exactly what is wrong with these arguments.

The PI also conflicts with the argument from neighboring theories, according to which since T has been supported by previously unconceived neighboring theories, it will continue to be supported by hitherto unconceived neighboring theories. This argument implies that T is true; however, the PI implies that it is false. We can resolve this conflict between the PI and the argument from neighboring theories by showing that either or both are incorrect. Again, I do not rule out the possibility that both are incorrect. It will also be interesting to see precisely what is wrong with these arguments.

The PI and the counter-induction for realism share the premise that the history of science is the graveyard of obsolete theories. However, they embody the uniformity and disuniformity principles, respectively. As a result, they conclude that current theories are false and true, respectively. They have opposing views about the list of obsolete theories. For pessimists, the longer the list is, the more plausible the conclusion of the PI is. For counter-inductionists, by contrast, the longer the list is, the more plausible the conclusion of the counter-induction is. This conflict between the PI and the counter-induction would be resolved if the PI and/or the counter-induction were shown to be incorrect.

Clearly, induction works in some cases, while counter-induction works in others. A fascinating future research project would specify the conditions under which induction works and those for when counter-induction does. My initial proposal is that we can choose counter-induction over induction when there is an improvement.

Suppose, for example, that there are successive attempts: $a_1$, $a_2$, and $a_3$. It turned out that $a_1$ and $a_2$ were failures. However, if $a_2$ is better than $a_1$, and $a_3$ is better than $a_2$, we can choose counter-induction over induction regarding $a_3$.

Many readers would not be able to resist the temptation to present some cases in which counter-induction does not work although there is an improvement. The existence of such cases, however, is compatible with my proposal that we can choose counter-induction over induction when there is an improvement. I am aware that even if improvement inheres in a series of failures, the improvement might not be significant enough to reach success. Keep in mind that I do not make the strong claim that improvement guarantees success.

My opponents would also feel tempted to present some instances in which success does not come no matter how many times we fail. For example, Isaac Newton could not transform copper into gold, no matter how many times he tried. However, counter-inductionists admit that there are some cases in which we cannot have success, no matter how many times we fail. No counter-inductionist would make the strong claim that in every case, if you fail a lot, you are likely to succeed or guaranteed to succeed. Counter-inductionists make the modest claim that in some instances, success will come if we fail enough.

The same is true of induction. It is crude to say against induction that in some instances, we succeed after some failures. For example, Ignaz Semmelweis succeeded after having several failures, as we noted in Sect. 2 of Chap. 8. Inductionists admit that there are some cases in which we succeed after some failures. No inductionist would make the strong claim that in every case, if you fail a lot, you are likely or guaranteed to fail indefinitely. Inductionists make only the modest claim that in some instances, if you have failed enough, you will fail indefinitely.

Exactly how many failures should scientists have before succeeding? Exactly how much improvement should be made before we are entitled to choose counter-induction over induction? I leave the task of answering these questions to prospective researchers. The accomplishment of this task will have a practical implication for funding agencies, viz., "it will tell funding agencies whether to keep funding a research team or not" (Park, 2017a, p. 222).

I ask again the prospective researchers on this topic to keep in mind that to reject counter-induction on the grounds that induction works in some cases is just as deplorable as to reject induction on the grounds that counter-induction works in other cases.

### *11.3.2 The Problem of Time Span*

The PI conflicts with the optimistic induction over realists that I enunciated in Sect. 2 of Chap. 4. This conflict raises the interesting issue of how long scientific theories need to persist for us to determine whether they are fated to be retained or discarded. This issue can also be discussed by future researchers of the realism debate.

The PI, making use of Larry Laudan's (1981, p. 33) list of superseded theories, holds that since the theories accepted before the twentieth century were overthrown, the theories accepted from the twentieth century onward will also be overthrown. By contrast, the optimistic induction implies that since the theories of the early twentieth century persisted for about one hundred years, they will last indefinitely into the future. These two inductions make conflicting predictions about the fate of current theories.

Pessimists would argue that even if the theories of the early twentieth century persisted for about one hundred years, it does not follow that they will survive indefinitely into the future. One hundred years is not long enough to run the optimistic induction, i.e., we need to watch theories for more than one hundred years before we can conclude that they are fated to remain unrefuted. In response, optimists would say that 100 years is long enough to construct the optimistic induction for them.

This confrontation between pessimists and optimists raises a problem that I call "the problem of time span." How long should we watch a theory before running a PI against it or an optimistic induction for it? In other words, what is the right time span for running a PI or an optimistic induction? The Ptolemaic theory lasted for about 1200 years (Fahrbach, 2011, pp. 141–142). This example suggests that we need to watch a theory for about 1200 years before running a PI against it or an optimistic induction for it. Consider, however, that at least three theories of light came and went in physics in the 120-year span from the early nineteenth century to the early twentieth century (Fahrbach, 2011, pp. 141–142). This example suggests that we should watch a theory for about forty years before running a PI against it or an optimistic induction for it.

There are many formulations of the PI in the literature. Only one of them refers to the time span, and its author is Putnam. He says, "Just as no term used in the science of more than fifty (or whatever) years ago referred, so it will turn out that no term used now (except maybe observational terms, if there are such) refers" (Putnam, 1978, p. 25). He might suggest that the appropriate time span is fifty years, so we should watch theories for about fifty years before running a PI against them or an optimistic induction for them. However, it is not clear why fifty years is the right time span.

Let me add that Putnam's PI is incorrect in light of the optimistic induction over realists (Park, 2021a). The optimistic induction over realists implies that most theoretical terms from the early twentieth century still refer in the twenty-first century. For example, "gravitational waves" and "oxygen" referred in the early twentieth century, and they still refer in the twenty-first century. Putnam contended in 1978 that the theoretical terms of the early twentieth century turned out not to refer and that the theoretical terms of his time will turn out not to refer in the early twenty-first century. Both of these claims are implausible.

### 11.3.3 The Problem of Selection

Antirealists make use of scientific ideas for their philosophical purposes. For example, Bas van Fraassen and Brad Wray use evolutionary theory to give an antirealist account of the success of science, as noted in Sect. 6 of Chap. 1. Thomas Nickles appeals to a psychological hypothesis to discredit realism, as noted in Sect. 7 of Chap. 1. Thomas Kuhn, Wray, and Nickles use evolutionary theory to flesh out their pessimistic account of scientific development, as noted in Subsection 3.2 of Chap. 2. Stanford appeals to the geological debate to innovate the realism debate, as noted in Sect. 11.4 of this chapter. However, antirealism bars these philosophers from making use of scientific ideas.

How can antirealists get around this objection? They might suggest that they can believe some claims of science without giving up antirealism, offering criteria for distinguishing between the claims of science that they can believe and the claims of science that they cannot. They might argue, for example, that they can believe evolutionary theory although they cannot believe quantum mechanics, without having to give up antirealism. Or they might argue that they can believe some parts of evolutionary theory although they cannot believe other parts of evolutionary theory, without giving up antirealism. I leave the task of fleshing out and defending this line of thought to antirealists.

Antirealists who aim to undertake the preceding task are asked to answer the following questions. Why does evolutionary theory belong to the group of theories that they can believe? Why does quantum mechanics not? What is the relevant difference between evolutionary theory and quantum mechanics? Alternatively, which components of evolutionary theory can they believe? Which parts of evolutionary theory can they not? What is the relevant difference between the parts of evolutionary theory that they can believe and the components of evolutionary theory that they cannot?

These questions give rise to what I call "the problem of selection," which is the problem of determining which claims of science are believable from antirealists' perspective. Only after antirealists solve this problem can they invoke some scientific ideas to defend their philosophical position, according to which the claims of science that they do not select are not believable. Otherwise, they are vulnerable to my previous charge that it is incoherent for antirealists to appeal to science.

Antirealists traditionally have suggested that the observational claims of science are trustworthy, whereas the theoretical claims of science are not. Based on this traditional suggestion, antirealists might argue that they can utilize the observational claims of science to justify their philosophical claims. The problems with this suggestion are that there is no strict dichotomy between observational and theoretical entities (Maxwell, 1962) and that the observational claims of distant past and future events might be epistemically just as remote as the theoretical claims of present events. Moreover, as the example of the geological debate between uniformitarians and catastrophists illustrates, the observational claims of distant past and future events, which are epistemically just as remote as the theoretical claims of present events, might be similar to the philosophical claims that antirealists wish to establish.

It is not clear how antirealists can solve the problem of selection. They should be conservative in their selection of scientific ideas. If they are liberal, they run the risk of shortening the distance between realism and antirealism. If there is only a narrow gap between realism and antirealism, there cannot be much debate between the advocates of the opposing positions, as we discussed in Sect. 3 of Chap. 6. It would be a challenge for antirealists to simultaneously select the believable claims of science and to keep in place the distinction between realism and antirealism.

### *11.3.4 The Unvirtuous Prediction*

Astronomers predict that the next solar eclipse will occur on a specific date in a particular place. Pessimists and selective realists predict that revolutions will happen in science; unlike astronomers, however, pessimists and selective realists do not specify when the anticipated events will take place. They do not select the year, for example, when quantum mechanics will be overthrown. For this reason, their prediction is unvirtuous. A task ahead for pessimists and selective realists is to specify the specific times when scientific revolutions will occur.

A prediction should meet the following conditions to be virtuous. First, it should specify the time and the place of the predicted event (Park, 2021a, pp. 582–583). Astronomers' prediction above meets this condition, whereas pessimists and selective realists' prediction does not. Second, a prediction should mention a high degree of probability. Pessimists and selective realists' prediction meets this condition, given that it asserts that the predicted events are likely to occur. Third, a prediction should be true. It is an open case whether pessimists and selective realists' prediction meets this condition.

The term "virtuous" is vague, and thus we cannot make a strict distinction between virtuous and unvirtuous predictions. It appears, therefore, that we cannot establish that pessimists and selective realists' prediction is unvirtuous. However, suppose that fortune-tellers predict that a mishap will happen to you. They do not specify the time of the predicted event, giving you no idea when you will be able to praise them for having been right or blame them for having been wrong. As a result, you regard fortune-tellers' prediction as unvirtuous (Park, 2021a, p. 585). If you do so, you should also consider pessimists and selective realists' prediction to be unvirtuous. This problem with pessimism and selective realism can be placed on the list of problems with the classic PI in Subsection 4.1 of Chap. 2, as well as the list of problems with the SI in Subsection 6.4 of Chap. 5.

Once pessimists and selective realists specify the specific times when current theories will be thrown out, we will have a better idea when we will praise them for having made true predictions or blame them for having made false ones. We will also have an idea when we will be able to praise selective realists for being *wisely* cautious or blame them for being *overly* cautious, i.e., when we will be able to praise them for believing only some assumptions of current theories or blame them for believing only some assumptions of current theories when they could have believed all or most assumptions of current theories. Unless pessimists and selective realists specify the specific times, we have no idea when we will be able to praise or blame them (Park, 2021a, p. 585).

Scientists strive for virtuous predictions. They do not rest content, for example, with the unvirtuous prediction that all humans are mortal. They instead do research to specify the times when each human will die. Molecular biologists can now roughly predict when you will die a natural death by investigating the length of the telomeres, the ends of your chromosomes. They can predict, for example, that you will die at around the age of ninety-five. In this spirit, pessimists and selective realists should not rest content with the unvirtuous prediction that current scientific theories are mortal (Park, 2021a, p. 590). They are encouraged to specify the specific times when current theories will die. Until they mention the specific times, their prediction will remain unvirtuous.

This criticism against the PI does not apply to the optimistic induction over realists, which holds that since the realists of the early twentieth century were right to believe the theories of their time, their philosophical offspring are also right to believe the theories of their times. This argument does not predict that current theories will be overturned. Therefore, it is not saddled with the burden of specifying the specific times when scientific revolutions will occur. However, blame will be put on optimists when current theories are rejected. Hence, the optimistic induction over realists runs more risk than the PI.

### *11.3.5 The Problem of Scope*

Consider globalism and localism in the realism debate that were introduced in Sect. 3 of Chap. 4. According to globalism, we should group together all theories in science and evaluate them collectively. Both the NMA and the PI exemplify globalism. The NMA pulls together all successful theories in all fields of science, claiming that most of them are true. The PI also pulls together all successful theories in all fields of past science, claiming that since most of them were false, most current theories in all fields of current science are false. As far as the NMA and the PI are concerned, all fields of science considered together are a legitimate scope of evaluation.

Globalism is rejected by such philosophers as Marc Lange (2002), Samuel Ruhmkorff (2013), Brad Wray (2016), and Jamin Asay (2019). They defend the position called "localism." According to localism, we should evaluate science on local scales. For Asay (2019), legitimate local scales are a discipline, a subdiscipline, and a single thesis, such as "species evolve." For example, it is right to evaluate together all theories in physics, but not all theories in physics and biology. Ruhmkorff (2013) and Asay (2019, p. 605) think that a local PI can be correct. A local PI ranges over a particular field of science, but not over all fields of science.

Globalism and localism are different responses to what might be called "the problem of scope." How should we set the scope of evaluation? How should we determine the boundary between the targets and the non-targets of our assessment? In general, our attitude toward a target may vary greatly, depending on whether it falls inside or outside the boundary. For example, our attitude toward a government official may vary significantly, depending on whether she falls within the scope of the evaluation in the next election. Similarly, our attitude toward a scientific theory may vary considerably, depending on whether it falls within the scope of a PI. Therefore, the problem of scope deserves special attention from the participants in the realism debate.

Localism does not, by itself, provide a solution to the problem of scope. Its problem can be brought to the fore with the discussion of the following two cases. First, many contemporary scientists are engaged in interdisciplinary research. What if physicists and biologists collaborate with each other under a common research agenda and develop a new theory? Should their new theory be bound together with physical or biological theories? If it were tied together with physical theories, it would be vulnerable to a local PI over physical theories; if it were bound together with biological theories, it would not. Thus, it falls victim to the PI over physical theories or not, depending on whether it is classified as a physical theory or a biological one.

Second, suppose that several theories were discarded in an old field of science, that scientists pioneer a new field in which they propose a new theory, and that the new field can be categorized as a subfield of the old field or as a separate field. If it were classified as a subfield of the old field, the new theory would be subject to the local PI that since several theories in the old field were discarded, it would also be discarded. By contrast, if the new field were classified as a separate field, the new theory would not follow suit. In sum, a scientific theory is subject to a local PI or not, depending on how we categorize its field of science.

These two cases indicate that our classification system determines our beliefs about whether a scientific theory will be falsified. Why is this a problem? It is not clear whether there is a rational basis for our classification system. Why should we classify the aforementioned new theory in the interdisciplinary field as a physical theory, but not as a biological theory? Why should we classify the aforementioned new field as a subfield of the old field, but not as an entirely separate field? Unless there are plausible answers to these questions, we can gerrymander the scope of the local PI. Gerrymandering the scope can result in the acceptance of my theory and the rejection of your theory, to my joy and to your disappointment. Therefore, there should first be a principled way to set the scope for a local PI. No local PI would be cogent unless its scope is first set justifiably. Localists, running a local PI, should be ready to answer the following questions. What is the scope of your local PI? Why did you fix the border of the scope as you did?

Lange (2002), Ruhmkorff (2013), Wray (2016), and Asay (2019) seem to have a theoretical resource to answer this question. They argue that scientific theories in different disciplines are under different epistemic circumstances. For example, the degree to which the existence of mitochondria is confirmed is different from the degree to which the existence of the Higgs field is confirmed. Therefore, it is wrong to put together all scientific disciplines and evaluate them collectively. Let me call this argument "the argument from epistemic homogeneity." It holds that we can assess together a set of epistemically homogeneous theories, i.e., we can assess a set of theories that are supported to a similar degree. All theories in physics or all theories in a subfield of physics are epistemically homogeneous, but physical and biological theories are epistemically heterogeneous. Hence, we can hold together all theories in physics or all theories in a subfield of physics, but we must not hold together physical and biological theories.

How about the new theory mentioned above from the interdisciplinary field of physicists and biologists? Adherents of the argument from epistemic homogeneity would say that this new theory can be drawn together either with physical or with biological theories, depending on whether it is epistemically homogeneous with physical or biological theories. If it is homogeneous with none of them, it should be drawn together with none of them, and it should be evaluated independently of physical and biological theories. The same goes for the new theory in the new field. The new field can be categorized as a subfield of the old field or a separate field, depending on whether the new field is epistemically homogeneous or heterogeneous with the old field.

However, there is a problem with the argument from epistemic homogeneity. If we can compare the degrees to which theories in different fields of current science are supported, we should also be able to compare the degrees to which a past theory and its present successor are supported. If we have such an ability, we would know that the present theory is better supported than the past theory and that the local PI from the past to the present theories is fallacious. In short, the argument from epistemic homogeneity is at variance with a local PI.

Pessimists might now run what I call "the argument from successorship." It holds that we can evaluate together a group of theories as far as they succeed one another. For example, the oxygen theory is the successor to the phlogiston theory, and thus a successorship exists between them. The same goes for the humoral and miasma theories. By contrast, the oxygen theory is not the successor to the humoral theory, and thus a successorship does not exist between them. Therefore, a local PI from the phlogiston to oxygen theories is legitimate, whereas a local PI from the humoral to oxygen theories is illegitimate.

A problem with the argument from successorship is that it is not clear why successorship should have any epistemic implications. Why is it right to say that because the phlogiston theory was false, the oxygen theory is false too, but wrong to say that because the humoral theory was false, the oxygen theory is false too? It is an implausible answer that the scientists of the phlogiston theory and those of the oxygen theory are in the same family line, and as a result, the former inherited *blunder genes* from the latter. By contrast, the scientists of the humoral theory and those of the oxygen theory belong to different family lines, and as a result, the former did not inherit blunder genes from the latter (Park, 2017b, p. 33). Those sympathetic to the argument from successorship are asked to give an account of why successorship acts like a pipeline that transmits falsity from old theories to new ones.

Seungbae Park (2019, pp. 362–365) defends the position called "scientific individualism," which asserts that an individual theory is a legitimate scope of evaluation. We should not group together two theories of current science, even if they belong to the same discipline. Neither should we group together a past theory and its present successor, even if they belong to the same subdiscipline. We should always evaluate a scientific theory solely on its merits as found in scientists' arguments for and against it. Scientific individualism is incompatible with both a local PI and a local SI.

Scientific individualism contrasts with scientific collectivism (Park, 2019, pp. 362–365). Scientific collectivism claims that we can appraise two or more scientific theories together. Unlike scientific individualism, scientific collectivism is compatible with a local PI and a local SI. It is also compatible with the suggestion that we can evaluate physical and biological theories together and with the suggestion that we can evaluate all theories in current science together. Asay's arguments against globalism apply to scientific collectivism (Park, 2019, pp. 365–377).

I must admit, though, that scientific individualism is not a solution to the problem of scope. Scientific individualism goes against a certain philosophical spirit. Philosophers usually group together different objects and strive to find their common properties. For example, moral philosophers bind together moral actions and attempt to find out what they have in common. They do not stop having only determined that a particular action is moral or immoral. For example, utilitarians suggest that moral actions have the common property of maximizing happiness. This shared property, once made visible to us, leads us to an insight about moral actions. Consider also that lawmakers do not enumerate all of the actions that are punishable under a specific clause in the criminal law; they only specify a common property of the actions that fall under the clause. This shared property, once made visible to us, proves to be convenient. For these reasons, lawmakers evaluate together a group of actions. Perhaps this point applies to scientific theories.

The problem of scope arises whether the target of evaluation is scientific theories or PIs. In Sect. 4 of Chap. 3, I argued that the PI over antirealists is the first idea in a separate domain, so it enjoys the first-mover advantage, i.e., it is not vulnerable to the grand PI that since the classic PI had infinitely many hidden problems, its successors also have infinitely many hidden problems. To argue so means that the PI over antirealists falls outside the scope of the grand PI, i.e., it is wrong to bind together the PI over antirealists, say, with the PI over realists, and then to claim that since the classic PI is incorrect, both the PI over antirealists and the PI over realists are also incorrect.

However, my opponents might group together the PI over antirealists and the PI over realists, claiming that since the classic PI is incorrect, both the PI over antirealists and the PI over realists are also incorrect. The basis for grouping together the PI over antirealists and the PI over realists is that they are both about philosophers. On my opponents' account, what matters is whether the target of a PI is philosophers or scientists. Both the PI over antirealists and the PI over realists are directed at philosophers; therefore, they belong to the same domain, and they can be bound together.

On my account, however, what matters is whether a PI supports or undermines realism. All first-order PIs that support realism are homogeneous, belong to the same domain, and can be brought together. All first-order PIs that undermine realism are also homogeneous, belong to the same domain, and can be grouped together. By contrast, the former PIs and the latter PIs are heterogeneous, belong to different domains, and should not be grouped together. The grand PI applies to the latter, but not to the former.

Note that the PI over antirealists falls prey to the grand PI or not, depending on how we set the scope for the grand PI. It is vulnerable to the grand PI once we set the scope of the grand PI as the domain of the PIs aimed at philosophers. By contrast, it is not susceptible to the grand PI once we set the scope for the grand PI as the domain of the first-order PIs that undermine realism. It is not clear which is the correct way to set the scope for the grand PI. In the absence of a principled way to set the scope, we can gerrymander the scope of the grand PI.

How should we set the scope of evaluation? This question needs to be answered by the participants in the realism debate whether they are realists or pessimists.

### 11.3.6 The Argument from Double Spaces

Van Fraassen (1989, p. 143) constructs the famous argument against realism called "the argument from a bad lot," and several philosophers invoke it to reject realism, as we saw in Sect. 5 of Chap. 1. In response to this situation, I construct a similar argument called "the argument from double spaces" (Park, 2020, 2021b). An interesting task for future researchers of these two arguments is to show that the argument from a bad lot is strong while the argument from double spaces is weak.

The argument from a bad lot holds that even if T is more successful than its conceived rival theories, we cannot believe that it is true because there might be unconceived rival theories that are better than it, i.e., because T "may well be the best of a bad lot" (van Fraassen, 1989, p. 143). In a nutshell, the argument from a bad lot asks realists to provide evidence that T is better than its unconceived rival theories, which is a demanding request.

The argument from double spaces starts from the distinction between the appearance and the reality of T. Suppose that T is successful, which means that it appears to be true and empirically adequate (EA). Although T appears so, it might be false and empirically inadequate. To put it differently, its appearance may be contrary to its reality. In addition, the argument from double spaces distinguishes between the two logical spaces of unconceived scientific theories. One is named as "the T-space" and the other as "the O-space." The T-space is so named because the reality of a theory selected from it is *transparent* to us, i.e., because its appearance and reality always agree with each other. By contrast, the O-space is so named because the reality of a theory selected from it is *opaque* to us, i.e., because its appearance and reality may agree or disagree with each other. The argument from double spaces is so named because it invokes the two logical spaces. It holds that we do not know whether T is selected from the T-space or from the O-space, so we do not know whether T has a certain semantic property, be it truth, empirical adequacy, and so forth (Park, 2021b, p. 4).

The argument from double spaces can be applied to realists and empiricists. Realists observe that T is successful, i.e., it appears to be true to them, so they conclude that it is true. The argument from double spaces says that realists' conclusion requires evidence that T is more likely to have been selected from the T-space than from the O-space. Empiricists observe that T is successful, i.e., it appears to be EA to them, so they conclude that it is EA. The argument from double spaces says that empiricists' conclusion also requires evidence that T is more likely to have been selected from the T-space than from the O-space (Park, 2020, pp. 64–66).

The argument from double spaces can also be directed at pessimists who argue that past theories were sent to the graveyard of dead theories, so current theories will also be sent to the graveyard. Current theories appear to be false to pessimists, so they conclude that current theories are false. The argument from double spaces says that pessimists' conclusion requires evidence that current theories are more likely to have been selected from the T-space than from the O-space.

T falls within the range of the argument from double spaces, whether T is a scientific theory or a philosophical theory. Van Fraassen's contextual theory, for example, falls within the range of the argument from double spaces. Van Fraassen claims that his contextual theory can explain the explanatory phenomena in science called "rejections" and "asymmetries" and then concludes that his theory is correct (van Fraassen, 1980, p. 112, pp. 130–134). The argument from double spaces says that his inference requires evidence that the contextual theory is more likely to have been selected from the T-space than from the O-space (Park, 2021b, p. 5).

In Chap. 5, we discussed five variants of putative realism: entity realism, minimal realism, axiological realism, van Fraassen's definition of scientific realism, and selective realism. They all fall prey to the argument from double spaces just like the contextual theory. Let me use entity realism as an example. Entity realists believe that oxygen is real on the grounds that we can manipulate it. The argument from double spaces says that entity realists' inference requires evidence that entity realism is more likely to have been selected from the T-space than from the O-space.

There are some differences between the argument from a bat lot and the argument from double spaces. The argument from a bad lot is directed at realism, requiring evidence that T is better than its unconceived rival theories. By contrast, the argument from double spaces is directed not only at realism but also at other positions that are less committed to the claims of science than realism, requiring evidence that T is more likely to have been selected from the T-space than the O-space. In sum, these two arguments ask the following questions, respectively: "Is T better than unconceived rivals?" and "Is T selected from the T-space?" (Park, 2020, p. 67). I ask future researchers to specify the reasons for thinking that the first question is legitimate, but the second question is not.

## 11.4 Summary

Stanford puts forward two proposals about how to renew the realism debate. The first is that rival participants stop debating whether our best theories are approximately true. The second is that the realism debate should be modeled on the geological debate. Concerning the first proposal, I object that it overlooks the fact that besides pessimists and selective realists, there are realists who deny that current theories will fall prey to scientific revolutions. Regarding the second proposal, I object that Stanford is pessimistic about scientific theories, including catastrophism and uniformitarianism, so it is self-defeating for him to make use of them.

However, it is a legitimate topic for realists and antirealists whether scientific revolutions will eventually occur. Beyond this topic, there are others in this vibrant philosophical field: How can we resolve the conflict of the PI with the argument from undiscovered evidence, with the argument from neighboring theories, and with the counter-induction for realism? Under what conditions should we choose induction over counter-induction and vice versa? How can we solve the problem of time span and the problem of selection? In what years will alternatives supersede current theories? How can we solve the problem of scope? Is the argument from double spaces weak, while the argument from a bad lot is strong?

## References

Asay, J. (2019). Going local: A defense of methodological localism about scientific realism. *Synthese, 196*(2), 587–609.

Fahrbach, L. (2011). How the growth of science ends theory change. *Synthese, 180*(2), 139–155.

Lange, M. (2002). Baseball, pessimistic inductions and the turnover fallacy. *Analysis, 62*(4), 2881–2885.

Laudan, L. (1981). A confutation of convergent realism. *Philosophy of Science, 48*(1), 19–49.

Maxwell, G. (1962). The ontological status of theoretical entities. In H. Feigl & G. Maxwell (Eds.), *Minnesota studies in the philosophy of science III* (pp. 181–192). University of Minnesota Press.

Park, S. (2017a). The uniformity principle vs. the disuniformity principle. *Acta Analytica, 32*(2), 213–222.

Park, S. (2017b). Scientific antirealists have set fire to their own houses. *PRO, 16*(1), 23–37.

Park, S. (2018). The pessimistic induction and the Golden rule. *PRO, 93*, 70–80.

Park, S. (2019). Localism vs. individualism for the scientific realism debate. *Philosophical Papers, 48*(3), 359–377.

Park, S. (2020). The appearance and the reality of a scientific theory. *Social Epistemology Review and Reply Collective, 9*(11), 59–69.

Park, S. (2021a). The unvirtuous prediction of the pessimistic induction. *Filozofia, 76*(8), 581–595.

Park, S. (2021b). On the argument from double spaces: A reply to Moti Mizrahi. *Social Epistemology Review and Reply Collective, 10*(2), 1–6.

Putnam, H. (1978). *Meaning and the moral sciences*. Routledge & K. Paul.

Ruhmkorff, S. (2013). Global and local pessimistic meta-inductions. *International Studies in the Philosophy of Science, 27*(4), 409–428.

Stanford, K. (2006). *Exceeding our grasp: Science, history, and the problem of unconceived alternatives*. Oxford University Press.

Stanford, K. (2015). Catastrophism, uniformitarianism, and a scientific realism debate that makes a difference. *Philosophy of Science, 82*(5), 867–878.

Stanford, K. (2018). A fond farewell to 'approximate truth'? *Spontaneous Generations: A Journal for the History and Philosophy of Science, 9*(1), 78–81.

van Fraassen, B. (1980). *The scientific image*. Oxford University Press.

van Fraassen, B. (1989). *Laws and symmetry*. Oxford University Press.

Wray, B. (2016). Method and continuity in science. *Journal for General Philosophy of Science, 47*(2), 363–375.

# Index

**A**
Abstract objects, 178, 183, 189, 196
Abstract world, 178–183, 189, 190, 193–197
Achinstein, P., 79
Akins, K., 12
Alai, M., 32, 33, 72, 76, 81, 92
Anti-epistemism, 150
Approximate truth, 3, 146, 222
Argument from a bad lot, 8, 9, 48, 58, 81, 109–111
Argument from double spaces, 236, 237
Argument from language, 179
Argument from neighboring theories, 68, 74–76, 83–85, 164, 209, 210, 227, 238
Argument from undiscovered evidence, 68, 71–74, 83, 84, 96, 209, 227, 238
Aristotle, 226
Asay, J., 24, 73, 104, 106, 107, 232, 233, 235
Attributive empiricism, 60, 99, 100, 118, 122, 126
Attributive realism, 89, 98–100, 118, 119, 122, 125, 126, 130, 131, 238
Axiological empiricism, 60, 96
Axiological realism, 90, 96–99, 237

**B**
Baker, A., 188
Balaguer, M., 196, 197
Beddor, B., 6
Benacerraf, P., 184, 187
Bird, A., 96, 161, 162, 171, 173
Bryant, A., 62
Busch, J., 187, 188
Byrant, A., 62

**C**
Callard, B., 182, 184–186
Carroll, S., 47
Causal platonism, 178, 184–186, 197
Chakravartty, A., 89
Chang, H., 96, 207
Classic pessimistic induction, 23–42
Cohen, J., 137
Colyvan, M., 186, 192
Concrete objects, 178, 183, 196
Concrete world, 178, 181, 183, 184, 187, 189, 190, 194, 195
Conditional empiricism, 122–124, 126, 128
Conditional realism, 116, 122–128, 131
Confirmational holism, 187, 188
Constructive empiricism, 11, 15, 46, 60–62, 64, 96, 99, 117, 119, 120
Contextual theory, 9, 55–58, 91, 92, 217, 237
Cordero, A., 101
Counter-induction for realism, 68, 76, 78, 79, 84, 85, 95, 149, 164, 208, 209, 227, 238
Cummins, R., 97, 98
Cuvier, G., 37

**D**
Darwin, C., 2, 8, 11, 12, 37, 71, 97
Dawes, G., 50
Dellsén, F., 24, 50, 116, 122, 123, 131, 132, 135–140, 142, 143, 145, 146, 151, 155, 159, 161, 171, 173, 208
de Regt, H., 135, 146–148, 151
Descartes, R., 226
Douglas, R., 202–204
Doxastic conditionalism, 115, 128–130


S. Park, *Embracing Scientific Realism*, Synthese Library 445,
https://doi.org/10.1007/978-3-030-87813-9

**E**
Einstein, A., 35, 36, 41, 142–144
Elgin, C., 135, 136, 146, 147, 151
Empiricism, 15, 16, 18, 19, 59, 110, 140, 171–174
End of history illusion, 12, 78
Enfield, P., 79
English notion of rationality, 49, 68, 79–84, 120–122
Entity realism, 90–93, 95, 109, 179, 237
Epistemic account, 155, 161–163, 169, 170, 172–174
Epistemic egoists, 57
Epistemic empiricism, 15, 59, 122, 124, 126
Epistemic imperative, 47, 54, 55, 84, 129, 130, 166
Epistemic pessimistic induction, 34, 39, 40, 48
Epistemic realism, 1, 4, 6, 14, 15, 20, 98–100, 118, 119, 122, 124–126, 130
Epistemic reciprocalism, 46, 56–58
Epistemism, 135, 139, 146, 148–150
Epistemological debate, 116, 118, 119
Evidential account, 163–171, 174

**F**
Fahrbach, L., 28, 47
Feynman, R., 206
Fictional understanding, 139, 140, 151, 161
Field, H., 187
Fine, A., 2, 9–11, 16, 20, 89, 111, 215
Fitzpatrick, S., 79
Friedman, M., 55, 56
Frost-Arnold, G., 2, 13–15, 20, 35, 40, 84

**G**
Galileo, G., 59, 69
Gap antirealists, 62, 63
Giere, R., 147, 148
Globalism, 73, 232, 235
Global pessimistic induction, 73
Global selective induction, 105
Golden Rule, 57, 84, 92, 225
Goldman, A., 6
Grand optimistic induction, 68, 83, 85, 108
Grand pessimistic induction, 24, 28, 31–38, 40–42, 48, 49, 58, 83, 108, 111, 235, 236
Grimm, S., 135

**H**
Hacking, I., 90–92
Healey, R., 47, 50, 56, 61, 117, 217
Hempel, C., 17, 19, 150
Historical inductions, 202–204
Humean argument, 23, 25, 26, 34, 150
Hume, D., 25, 34, 63

**I**
Indispensability argument, 178, 186–193, 197
Inference to the best explanation (IBE), 2, 8–11, 13, 20, 23, 26, 57, 75, 76, 80, 84, 90–93, 121, 217
Instrumentalism, 90, 94, 128, 129, 212–214, 216–218
Interactive theoretical entities, 191, 198

**K**
Kant, I., 55
Khalifa, K., 8
Kitcher, P., 104, 158–160, 171
Kuhn, T., 27, 28, 61–63, 117, 158, 170, 180, 204, 230
Kukla, A., 16, 215

**L**
Ladyman, J., 8, 10, 15, 24, 49, 101
Lamarck, 37
Lange, M., 232, 233
Laudan, L., 2, 5–7, 9–11, 15, 20, 23–26, 28, 29, 41, 77, 78, 96, 103, 125, 127, 158, 171, 229
Laudan's objection, 2, 5, 6, 23, 24, 96
Lavoisier, A., 37, 55
Leplin, J., 15, 16, 18, 19, 117, 215
Lipton, P., 79, 104
Localism, 73, 232
Local pessimistic induction, 73, 105, 232–235
Local selective induction, 105, 106, 235
Long, D., 216
Lyons, T., 4, 6, 16–18, 20, 46, 47, 78, 96, 98, 117, 211, 215

**M**
Mach, E., 54
Maddy, P., 187, 188
Markram, H., 202–204
Mathematical convenientism, 192–193

Mathematical fictionalism, 178, 196, 197
Mathematical inferentialism, 178, 193–197
Mathematical realism, 178, 179, 181, 184, 186–188, 190, 192, 193, 195, 196, 198
Meta-inductions, 202–204
Middlism, 94
Minimal realism, 93–95, 109–111, 237
Mizrahi, M., 2, 14, 16, 19, 29, 47, 61, 84, 101, 110, 159, 215
Moore, G., 46, 50–52, 59, 62–64, 91
Moore's paradox, 46, 50–52, 59, 62–64, 91, 128, 136, 141, 144, 213
Morrison, M., 92
Müller, F., 72, 76, 81, 92
Musgrave, A., 4, 16, 90, 91, 117, 121, 215

**N**
Narrow formulation, 120, 132
Newton, I., 29, 35, 36, 41, 53, 228
Nickles, T., 2, 13, 14, 20, 26–28, 33, 68–70, 78, 121, 122, 208, 211, 230
Niiniluoto, I., 155, 162
Noetic account, 155, 161, 165, 168, 173, 174
No-miracles argument (NMA), 2–10, 13, 14, 20, 26, 46, 67, 75, 78–80, 83–85, 99, 116, 118, 120, 121, 151, 204, 222, 223, 232
Noncommittalism, 216

**O**
Optimistic induction over realists, 68, 69, 83, 84, 101, 108, 228, 232

**P**
Papineau, D., 119, 131
Park, S., 31, 34, 46–48, 56, 61, 94, 117, 211, 234
Parsons, C., 192, 193
Pessimistic induction (PI), 2, 11, 23–25, 36, 46, 76, 79, 85, 92, 94, 101, 109, 117, 118, 120, 121, 148, 150, 162, 172, 173, 187, 202, 203, 205, 207, 218, 221, 227, 228, 231, 232, 235, 238
Pessimistic induction against selective realism, 107, 109
Pessimistic induction over antirealist explanations, 46, 47, 211
Pessimistic induction over antirealists, 46, 48–50, 63, 95, 99, 232, 235, 236
Pessimistic induction over pessimists, 46–49, 63
Pessimistic induction over realists, 33, 34, 40, 48, 68, 83, 235
Poincaré, H., 24, 54
Predictive similarity, 19, 20, 46
Problem of deceptive speech acts, 46, 51, 52, 62, 91, 128, 136, 213
Problem of disconcerting questions, 46, 50–52, 62, 63, 91, 128, 136, 213
Problem of misleading evidence (PME), 35–41, 48, 83
Problem of scope, 232, 233, 235, 238
Problem of selection, 230, 231, 238
Problem of time span, 228, 229, 238
Problem of unconceived alternatives (PUA), 29, 30, 32–41, 48, 50, 59, 71, 72, 74, 75, 82–85
Problem of unobserved anomalies, 59, 61, 100
Problem-solving account, 155, 158–159, 166, 167, 171, 174
Prussian notion of rationality, 79, 121
Psillos, S., 2, 14, 27, 61, 104, 149, 160, 205, 217
Ptolemy, 165, 167, 169, 170
Putative realism, 89
Putnam, H., 2–6, 8, 24, 67, 93, 98, 116, 117, 119, 120, 132, 186, 204, 211, 229

**Q**
Quine, W., 182, 186, 187, 189, 190, 192

**R**
Relative pessimistic induction, 32, 33, 40, 48
Resnik, M., 186, 192
Richard Boyd, 3
Roush, S., 79
Ruhmkorff, S., 74, 232, 233

**S**
Saatsi, J., 93–95, 109
Sankey, H., 2, 160
Scientific understanding, 135, 139–141, 146, 147, 149, 152, 161
Seeman, J., 215
Selective induction (SI), 108, 202, 203, 205, 221
Selective realism, 27, 69, 90, 101–109, 111, 117, 149, 188, 205, 206, 223, 231, 237

Semantic account, 155, 159–160, 167, 168, 172, 174
Semantic realism, 1, 4, 6, 14, 20, 99, 100
Semmelweis, I., 156–163, 165–170, 174, 175, 228
Shaw, J., 201–204, 207
Stahl, 37
Stanford, K., 2, 16, 19, 20, 24, 25, 29–33, 37, 41, 48, 49, 59, 69–73, 75–77, 81, 82, 94, 102, 103, 149, 150, 205, 208, 212, 214, 215, 221–227, 230, 238
Surrealism, 2, 16, 17, 46, 214–218
Synthetically adequate, 194–196

**T**
Tricle, 181–182
*Tu Quoque*, 84, 85

**U**
Unvirtuous prediction, 231, 232

**V**
van Fraassen, B., 2, 7–9, 11, 12, 20, 26, 46, 48–50, 55–58, 60, 61, 79–81, 83–85, 91, 96, 98–100, 109, 116, 117, 119–123, 131, 132, 138–140, 158, 217, 230, 236, 237
Vickers, P., 24, 117

**W**
Weismann, 37
Wide formulation, 120
Winther, R., 50
Woodburne, M., 74
Worrall, J., 104
Wray, B., 8, 11, 12, 14, 20, 26–28, 33, 48, 59, 61, 68–71, 119–121, 159, 217, 230, 232, 233
Wright, L., 97

**Z**
Zinsmeister, W., 74

The manufacturer's authorised representative in the EU is Springer Nature Customer Service Centre GmbH, Europaplatz 3, 69115 Heidelberg, Germany. If you have any concerns regarding our products, please contact ProductSafety@springernature.com

Printed and bound by CPI Group (UK) Ltd, Croydon, CR0 4YY
30/09/2026
02225696-0013